The Protection of Historic Properties

The Protection of Historic Properties

A comparative study of administrative polices

Consuelo Olimpia Sanz Salla
Universidad Jaume I, Spain

WIT*PRESS* **Southampton, Boston**

Consuelo Olimpia Sanz Salla
Funcionaria Administración Local con Habilitación Nacional, en excedencia y Doctora en Derecho,
Profesora Titular de Escuela Universitaria de Derecho Administrativo,
Universidad Jaume I,
Castellón, Spain

Published by

WIT Press
Ashurst Lodge, Ashurst, Southampton, SO40 7AA, UK
Tel: 44 (0) 238 029 3223; Fax: 44 (0) 238 029 2853
E-Mail: witpress@witpress.com
http://www.witpress.com

For USA, Canada and Mexico

WIT Press
25 Bridge Street, Billerica, MA 01821, USA
Tel: 978 667 5841; Fax: 978 667 7582
E-Mail: infousa@witpress.com
http://www.witpress.com

British Library Cataloguing-in-Publication Data

A Catalogue record for this book is available
from the British Library

ISBN: 978-1-84564-404-8

Library of Congress Catalog Card Number: 2008940357

The texts of the papers in this volume were set individually by the authors or under their supervision.

Printed in Great Britain by Athenaeum Press Ltd.

For my husband Martin and
for my children, Martin and Giselle

Contents

Acknowledgements

This genesis of this publication was a suggestion from my thesis director, D.Ramón Martin Mateo, who felt that, on a national level, practical legal studies examining the direct and effective application of heritage law to industrial sites were few and far between. This led to an original study, in Spanish, on heritage law in my own country, applied to a specific case – the Gerencia in Puerto de Sagunto. I would like to express my gratitude to D. Ramón for the original suggestion, and for his continued support and words of kindness throughout my career.

I would also like to extend my gratitude to the members of the Public Law Department at Castellón University, excellently led by Ricardo Garcia Macho, and to my students, too many to name, who keep me on my toes.

Within the context of this publication, I would also like to thank Professor Carlos Brebbia of the Wessex Institute of Technology for his comments and suggestions on the book. It was principally as a result of his comments that I decided to extend the scope of the publication to include sections on English and North American law. I would like to extend my gratitude to Elizabeth, Rachel and Lorraine at WIT press for their patience with me at various moments in the publication process.

Many people have been kind to me when requesting material. In this sense I would like to thank Clotilde Luce from Uphold our Heritage, who sent images and comments on the Jackling House, Hanns-Wilhelm Grobe from the World Cultural Heritage Site at the Völklingen Ironworks – the European Centre for Art and Industrial Culture, Adrian Farmer who was kind enough to send images on the Derwent Valley site and to El Transbordador de Vizcaya, S.L. for the image of the Puente Colgante in Vizcaya. I would like to extend a special mention to Julian Negredo Sanchez, whose fabulous photographs of the wonderful industrial heritage in Sagunto inspired me, and who has deservedly received many national and international prizes for his work, including the Premio Colección in the XXXII Certamen de Fotografía Calella in Barcelona 2008, and the Gold medal in the 17th Trierenberg Super Circuit 2008. I would also like to thank Miguel Angel Piqueras for the photographs of Ojos Negros taken during his academic years. Also Ximo Revert for his help in understanding the history of the movement to save the Gerencia in Sagunto.

Special thanks go to my husband, Martin Green, whose help in writing this book is unquantifiable.

On the English side of my family, I would like to dedicate the book to Ron and Eileen, who have always been wonderfully kind to me, and to Peter, Mie, Christopher, Jane, Faisal, Shaz, Jamie and William.

On the Spanish side of my family, to my nieces, Sandra and Inma, and my sister Manolita. Also to Eliseo, and to Elena and Lidon and their families.

Finally, I would like to dedicate the book to my father, Julio who worked all of his life for his family until he could do so no longer, to my mother, Consuelo, whose heart of gold and incredible spirit in the face of adversity was always an inspiration to me, and, above all, to our fabulous children, Martin and Giselle, who have brought us love and joy, and who give meaning to all we do.

Preface

This publication is intended to be an introduction to the protection of historic properties by public agencies in three very different legal systems – the United States, the United Kingdom and Spain, as well as analysing international efforts to protect historic heritage. It is not intended as a comprehensive study of any of the legal systems, but rather as a tool to give the reader the chance to compare the legal techniques used in each different administrative system.

The book outlines the historical trends in each legal tradition, as well as present day regulation, using a multitude of examples of how historic buildings and heritage sites are protected in each country. While the publication is intended to cover legal mechanisms established to conserve all types of historic building, we will pay special attention to the protection given to industrial heritage, giving examples wherever possible of how a particular legal technique has been used to protect industrial buildings. We will also examine how case law has affected heritage protection, sometimes in unexpected ways.

An important part of this study will deal with statutes and legal techniques designed integrally to protect cultural heritage. However, important protection is also offered to historic buildings through the inclusion of cultural heritage protection into environmental statutes. This is especially important in the case of those statutes that deal with the preparation of Environmental Impact Statements. In this sense United States law is especially important. The regulatory system of statutes such as the National Environmental Policy Act, as well as the Section 106 process included in the National Historic Protection Act, enter into extensive detail on the responsibilities of public administration in locating, identifying and evaluating historic sites. The extensive case law interpreting these regulations is also an important tool in understanding the importance of Environmental Impact Assessments in the protection of historic properties.

The study will also deal with financial incentives to protect historic properties, and once again, United States regulation is important in our understanding of methods for protecting our Historic heritage.

As far as the layout of the study is concerned, we have chosen to separate each legislative system into distinctive sections in the book. We therefore deal first with relevant international regulation on the protection of historic properties. We then discuss protection of historic properties in separate chapters dealing with the law in the United States, the European Union, the United Kingdom and Spain.

Finally, this publication is intended as an academic study comparing specific aspects of the law that protects historic properties in the three countries previously mentioned, and follows a style commonly used in Spanish research works – while being primarily a study of legislation and jurisprudence, it also includes brief references to scholarly opinion that has helped to shape and interpret the law on historic properties, with the only intention of allowing the reader to delve further into this area of study by going directly to those authors' works. A publication of this nature cannot deal with all aspects of the subject matter in the level of depth that we would like. To do so would be to duplicate the academic research ably carried out by other scholars. I hope the reader who wishes to follow up on any particular area will find the authors cited and the footnote references a useful guide on where to go for further information.

While I have only made brief reference to their work, and have only included a small fraction of scholarly opinion on the subject, I hope that I have managed in all cases to reflect accurately the opinions of the authors cited, and would like to thank all those who, by way of their articles and publications, have created an important body of academic opinion which others, including the present author, can use as both a point of reference and a starting point for future studies.

Introduction

A comparative study on the protection of historic properties must first deliberate on the concept of heritage. Dictionary definitions are of no use here. Heritage is a concept in constant evolution that requires a wider study, and has been mulled over by a multitude of scholars.

A first step in defining heritage is underlining its shared nature. Heritage is something that unites and binds social groups. It is no use thinking of heritage in terms of individuals. The collective nature of heritage is crucial to our understanding of the object of protection. Something that is known to all members of a group – a painting, a significant building, a language, a work process that defines an age – is shared, and is part of that group's heritage. Something that, while an important artistic creation is known only to its owner or creator, is hardly a shared item, and though as artistically perfect that it may be, is not heritage. Of course, being shared is not enough to define something as heritage. Heritage is something that "bears witness" to previous eras, something that in an intangible way, defines previous civilizations, or earlier versions of our own.

So, heritage defines and gives meaning through its collective nature and by its capacity to bear witness. As Lowenthal points out, while many symbolic and historical locations are rarely visited, the past is integral to our sense of identity, and the ability to recall and identify with our own past gives existence meaning, purpose and value, to such an extent that "identification with earlier stages of one's life is crucial to both integrity and well being".[1] Evans examines how this shared concept defines heritage conservation law, which "can be seen as a creature of international law, municipal statute and local custom which relates inter alia to those aspects of a shared community inheritance that a community chooses to foster, protect and pass on to future

1 LOWENTHAL, D. The past is a foreign country, Cambridge University Press, Cambridge 1986.

generations, and to this end, is prepared to limit public and private rights in a wider public interest".[2] Nivala asserts that, although a building has physical boundaries, its meaning and value depend on its relationship to the city outside them. Not just the physical space around the building, but also the economic, social, political and historical forces converging on its site. The building provides a frame for examining those forces.[3]

A second way of approaching the concept of heritage is to assess its relative, constantly changing nature. If we part from the premise that heritage is shared, and defines and unites us as a group or society, then we have to accept as a logical consequence its changing nature. Crew and Sims argue that "in elevating or admiring one piece of the past we tend to ignore and devalue others. One reality lives at the expense of countless others".[4] For Schouten, heritage is a product, and as such, it is as subject to differences in validation and interpretation as the historical process itself. For this author, "heritage changes over time in the way it is presented and also in the ways in which the public reacts to its presentation".[5] Therefore, heritage as a historical reality owes its existence to an act of interpretation, subject to constant change.

This very relative nature of heritage could lead us to another definition, proposed by Rose, who argues that heritage can be defined as "any cultural possession which is actively passed from one generation to another and which is valued by the donor and recipient alike". For this author, the focus is upon that part of culture which is held to be so important that the act of transmission to the next generation is talked about and given high priority by society.[6]

This relative aspect to the concept of heritage is of course central to heritage law. A common thread to all legal frameworks to protect heritage is a necessary lack of concrete definition when placing a limit on what is, of course, the very heart of the question. What is to be protected, and what is to be abandoned? The ways we give meaning to things as a group is complex, and assigning meaning to them, as well as defining the relationship between material and culture, are for Colley, "questions which have long vexed anthropologists and archaeologists. The issue is complex because what material things mean to people depends on their perceptions, and these can and do

---

2 EVANS, M. Principles of Environmental Heritage and Law, Prospect, Sydney 2000, p. 18.

3 NIVALA, J. in "Saving the Spirit of Our Places: A View on Our Built Environment" 15 UCLA J. Environmental. Law & Policy 1.

4 CREW, S. and SIMS, J. "*Locating Authenticity: fragments of a dialogue*" in Exhibiting Cultures: the poetics and politics of a museum display. Washington DC 1991, p. 160.

5 SCHOUTEN, F. "*Heritage as a Historical Reality*" in Heritage Tourism and Society, ed. HERBERT, T., Pinter Press 1995.

6 ROSE, R. Artistic Heritage in a changing Pacific, University of Hawaii 1993, p. 225.

change".[7] All the legal systems analysed use a variation on the term "significant historical or cultural value", and as we have seen, that is an idea that is in constant evolution.

Giannini addresses the problem of defining heritage in a legal context. "The question that must be formulated is whether it is correct juridically to say of a 'thing' that it is a material witness to civilization, or whether, on the contrary, it is a prejuridic concept with only generic value". The author argues that the concept of a cultural object is a concept that can be empirical for the scientist who studies civilization and culture, but for the legal expert can only be an open concept which is given true meaning by other academic disciplines on a case-by-case basis. Only in this sense can heritage as a material witness to previous civilizations be accepted as a valid legal concept, as a concept that can only be correctly defined by academics from other disciplines.[8] The concept of heritage as a witness to the values of civilization makes it an ideal terminology that includes many different types of heritage, not only from the distant past, but also up to items from the present day.

Giannini argues that the existence of a value judgement in the administrative process is inevitable, and that this very value judgement is precisely the act that justifies the declaratory resolution of the item of heritage. The author points out that this value judgement is sometimes referred to as the "motivation" behind the administrative resolution, but that this is misleading, as "motivation" implies that there is a pondering of different interests, whereas the declaration simply states whether or not the qualities that constitute the "special interest" of the heritage exist or not. The declaration does not argue the case, either for or against the existence of "special interest". The declaration simply states that the "special interest" does, or does not, exist.[9]

Martín Mateo points out the worrying nature of a branch of legislation in which the object to be protected cannot be effectively defined in the legal text itself. The author points out that

> as can be easily deduced, legal protection of a given element of heritage depends on a technical valuation that has be incorporated into the legal system so that it can have legally binding consequences. As a result, there will be a resort to experts from other fields whose criteria must be given transcendence by public administration given that the rights of those individuals that own the item of heritage are greatly

7 COLLEY, S. Uncovering Australia, Archaeology, Indigenous People and the Public, Allen & Unwin, 2002, p. 33.

8 GIANNINI "*I benni culturali*" Rivista trimestrale di Diritto Público año 26, núm 1 1976.

9 GIANNINI "*I benni culturali*" op. cit.

affected, and that their rights are protected by the Constitution itself which protects their right to judicial review, making absolutely necessary as a first step the corresponding administrative act. This produces a large dose of legal uncertainty, which tends to terrify the legal scholar.[10]

Evans, for his part, stresses the need for clarity and coherence in the perception and application of this branch of law. Heritage decision making must be based on rational and justifiable principles. This is especially important to avoid being caught in the paradox that, while heritage plays a significant part in our "individual and collective self definition...we in turn define our own heritage".[11] Rand, for his part, warns that while "most would agree that at least some degree of historic and cultural preservation is an important part of a nation's heritage and culture...preservation and historic and cultural significance are such broad and ambiguous concepts, that laws regarding preservation are particularly susceptible to arbitrariness".[12]

A third way of defining heritage is as a product, particularly in the growing field of cultural tourism. Prentice points out that as heritage becomes more closely linked to tourism, so the number of heritage sites increase.[13] Indeed, heritage laws and planning place great emphasis on the potential of heritage tourism.

The study of the concept of heritage has been an important part of academic texts on the subject. We have pointed out three major elements of the term: its shared nature, its relative nature dependant on an interpretation of vague concepts, and its utility as a product for tourism and urban regeneration. As we will see throughout this book, legal systems have struggled to tie down the concept, and have adopted different mechanisms to decide what is to be protected by public institutions, and what is, effectively, to be abandoned from a public perspective.

---

10 MARTÍN MATEO, Prologue to the book by ALONSO IBÁÑEZ, *El patrimonio histórico. Destino público y valor cultural*, op. cit., pp. 23–24.

11 EVANS, M. Principles of Environmental Heritage and Law, Prospect, Sydney 2000, p. 1.

12 RAND, K. "Nothing lasts forever: toward a coherent theory in American preservation law" in 27 U. Mich. J.L. Reform 277, 1993.

13 PRENTICE, R. Tourism and Heritage places, Routledge 1993. See also BONIFACE P. and FOWLER P.J. 1993, Heritage and Tourism in the 'global village', London: Routledge.

Chapter 1

The Protection of Historic Properties in International Law

1 Historical Evolution of International Law Protecting Historic Properties

International law is a complex, plural and heterogeneous body of legal regulation which has established a series of minimum standards of behaviour that individual states, at least nominally, incorporate into their own legal systems.[14]

The notion of cultural heritage, in all its complexity, owes its existence first to the work of the League of Nations, and then to the United Nations. The international lawmaking process began with the Athens Charter of 1931 for the Restoration of Historic Monuments,[15] and was further developed after the foundation of UNESCO in 1945. Important milestones were the Venice Charter (international charter for the conservation and restoration of monuments and sights) of 1964 and the Paris Convention of 1972 concerning the protection of the World Cultural and Natural Heritage.

The Athens Charter of 1931, which as Rodwell points out "was the first document to set out the scientific principles for the preservation and restoration of Historic Monuments at an international level",[16] took into account new developments in thinking on heritage, especially the general

14 GEMMA PRIETO, M. "*Derecho Internacional y bienes culturales*", Patrimonio Cultural y Derecho **2**, 1998, p. 12.

15 The Athens Charter for the Restoration of Historic Monuments. Adopted at the First International Congress of Architects and Technicians of Historic Monuments, Athens 1931.

16 RODWELL, D. Conservation and Sustainability in Historic Cities, Blackwell Publishing, 2007, p. 12.

tendency which recognised community rights as a counterbalance to private ownership. It also acknowledged the difficulty of reconciling public law with the rights of individuals, and felt that sacrifices made by property owners in the general interest should be recognised. The Charter also recommended that public authorities in each country be empowered to take emergency protective measures in emergency situations.

The Charter also made recommendations on the aesthetic enhancement of monuments, on restoration techniques, and urged greater international cooperation to further protect monuments through the League of Nations.

The Venice Charter for The Conservation and Restoration of Monuments and Sites,[17] which continues to be used as a baseline document for modern heritage doctrine,[18] asserts that the concept of historic monuments embraces not only individual buildings, but also the urban or rural setting where evidence of a particular civilization, a significant development or an historic event are found. The charter extended the concept of heritage beyond great works of art to include more modest works that, over time, have acquired cultural significance. One of the more important ideas in the Charter was that, imbued with a message from the past, the historic monuments of generations of people remain to the present day as living witnesses of their age-old traditions, and so "the intention in conserving and restoring monuments is to safeguard them no less as works of art than as historical evidence".[19] It felt that a common responsibility to safeguard them for future generations exists and recognised a duty to hand them on in the full richness of their authenticity. To this end, it was deemed essential that the principles guiding the preservation and restoration of ancient buildings should be agreed and be laid down on an international basis, with each country being responsible for applying the plan within the framework of its own culture and traditions. The Charter stated that the aim of conservation and restoration work is to safeguard monuments both as works of art and as historic evidence, and that added components should be in harmony with the existing monument, but distinguishable from the original.

---

17 The Venice Charter: International Charter for the Conservation and restoration of Monuments and Sites. II International Congress of Architects and Historic Monument Technicians, Venice, 1964. Adopted by ICOMOS in 1965.

18 RODWELL, D. Conservation and Sustainability in Historic Cities op. cit. p. 12 See also LARKHAM, P. Conservation and the city, Routlege 1996, p. 266.

19 Article 3 Venice Charter.

2 The Paris Convention of 1972 Concerning the Protection of the World Cultural and Natural Heritage

The Paris Convention[20] which is, with 176 member countries, the most ratified of the UNESCO treaties,[21] went beyond these earlier efforts, uniting the two concepts of nature and heritage. The preamble of the convention pointed out that both cultural and natural heritage is increasingly threatened with destruction not only by the traditional causes of decay, but also by changing social and economic conditions. Francioni points out that the importance of this holistic approach to nature and culture has been revindicated in recent years, with a rediscovery of the vital connection between society and its natural surroundings, between traditional culture and the use of natural resources, and between economic activity and limits on the use of natural resources.[22]

The preamble goes on to argue that protection of this heritage at national level was often incomplete because of the scale of the resources required and of the insufficient economic, scientific, and technological resources of the country where the property to be protected is situated. In view of the magnitude and gravity of the new dangers threatening them, "it is incumbent on the international community as a whole to participate in the protection of the cultural and natural heritage of outstanding universal value, by the granting of collective assistance which, although not taking the place of action by the State concerned, will serve as an efficient complement thereto".[23]

2.1 Definition of Heritage in the Paris Convention

In order to achieve these ends, the convention affronts two challenges: a definition of world heritage and the establishment of a mechanism of international cooperation.

The convention divides world heritage into two distinct areas: cultural heritage (article 1) and natural heritage (article 2). Cultural heritage is divided into monuments (comprising architectural works, works of monumental sculpture and painting, elements or structures of

20 Convention Concerning the Protection of the World Cultural and Natural Heritage. Adopted by the General Conference at its seventeenth session, Paris, 16 November 1972.

21 FRANCIONI, F. "Treinta años después ¿esta la Convención sobre el Patrimonio Mundial preparada para el siglo XXI" in Patrimonio Cultural y Derecho 7, 2003 p. 12.

22 FRANCIONI op. cit. p. 15.

23 Preamble to the convention.

an archaeological nature, inscriptions, cave dwellings and combinations of features, which are of outstanding universal value from the point of view of history, art or science) groups of buildings (comprising of groups of separate or connected buildings which, because of their architecture, their homogeneity or their place in the landscape, are of outstanding universal value from the point of view of history, art or science) and sites (works of man or the combined works of nature and man, and areas including archaeological sites which are of outstanding universal value from the historic, aesthetic, ethnological or anthropological point of view).[24]

For its part, natural heritage is made up of the following three categories:

- natural features consisting of physical and biological formations or groups of such formations, which are of outstanding universal value from the aesthetic or scientific point of view;
- geological and physiographical formations and precisely delineated areas which constitute the habitat of threatened species of animals and plants of outstanding universal value from the point of view of science or conservation; and finally
- natural sites or precisely delineated natural areas of outstanding universal value from the point of view of science, conservation or natural beauty.[25]

Evans argues that the ethical rationale behind the concept of a universally shared heritage as developed by the World Heritage Convention transcends the restricted viewpoint of heritage as a highly selective reading of the past.[26] For this author, the idea of common heritage of mankind not only fuses cultural and natural heritage, but also ideas about the past, present and future. “In doing so, the ethical basis of the human relationship with the rest of the animate and inanimate biosphere shifts from one asserting sovereignty and dominance to one of cooperative sustainability and furthering a present and future common good”.[27]

---

24 Article 1 Convention of Paris.
25 Article 2 Convention of Paris.
26 See in this sense ANDERSON, M. “*In search of woman's public history: heritage and gender*” in Public History Review number 3.
27 EVANS, M. Principles of Environmental Heritage, Prospect, Sydney 2000, p. 19.

2.2 The Criteria Defining World Heritage

To be included in the definition of world heritage, a monument, group of buildings or site is evaluated according to a series of criteria decided by an Intergovernmental Committee for the Protection of the Cultural and Natural Heritage of Outstanding Universal Value, called "the World Heritage Committee".[28] Up until the end of 2004 there were six criteria for cultural heritage and four criteria for natural heritage. In 2005 they were united in ten criteria applicable to both categories. To be nominated as world heritage, the monument, site, or group of buildings must be of outstanding universal value and fulfil at least one of the following criteria:

1. To represent a masterpiece of human creative genius;
2. To exhibit an important interchange of human values, over a span of time or within a cultural area of the world, on developments in architecture or technology, monumental arts, town-planning or landscape design;
3. To bear a unique or at least exceptional testimony to a cultural tradition or to a civilization which is living or which has disappeared;
4. To be an outstanding example of a type of building, architectural or technological ensemble or landscape which illustrates significant stage(s) in human history;
5. To be an outstanding example of a traditional human settlement, land-use, or sea-use which is representative of a culture (or cultures), or human interaction with the environment especially when it has become vulnerable under the impact of irreversible change;
6. To be directly or tangibly associated with events or living traditions, with ideas, or with beliefs, with artistic and literary works of outstanding universal significance. (The Committee considers that this criterion should preferably be used in conjunction with other criteria);
7. To contain superlative natural phenomena or areas of exceptional natural beauty and aesthetic importance;
8. To be outstanding examples representing major stages of earth's history, including the record of life, significant on-going

28 Article 8 Convention of Paris.

geological processes in the development of landforms, or significant geomorphic or physiographic features;

9. To be outstanding examples representing significant on-going ecological and biological processes in the evolution and development of terrestrial, fresh water, coastal and marine ecosystems and communities of plants and animals; and finally
10. To contain the most important and significant natural habitats for in-site conservation of biological diversity, including those containing threatened species of outstanding universal value from the point of view of science or conservation.[29]

Properties also have to meet tests of "authenticity" and "integrity". Properties may be understood to meet the conditions of authenticity if their cultural value (as recognized in the nomination criteria proposed) are truthfully and credibly expressed through a variety of attributes including:

– form and design;
– materials and substance;
– use and function;
– traditions, techniques and management systems;
– location and setting;
– language, and other forms of intangible heritage;
– spirit and feeling and other internal and external factors.

The Committee recognises that attributes such as spirit and feeling do not lend themselves easily to practical applications of the conditions of authenticity, but nevertheless considers that they are important indicators of character and sense of place, for example, in communities maintaining tradition and cultural continuity.

Integrity is considered as a measure of the wholeness and intactness of the natural and/or cultural heritage and its attributes and involves assessing to what extent the property includes all elements necessary to express its outstanding universal value, the extent to which its size is adequate in order to ensure the complete representation of the features and processes which convey the property's significance, and the extent to which it suffers from adverse effects of development and/or neglect.

29 Operational guidelines for the implementation of the world heritage convention, WHC 05/2 2nd February 2005.

2.3 The World Heritage Fund

The international community, as represented by UNESCO, although establishing the world heritage list and fixing certain obligations insofar as conservation is concerned, is conscious that not all countries have sufficient resources to protect the heritage that has been included on the list, and, in this sense, the Paris Convention follows the principle of subsidiarity.[30] The international community has, therefore, an obligation to support, but not supplant or substitute independent countries in areas such as identification or rehabilitation of heritage sites, but rather is directed towards tasks such as the prevision of expert advisory bodies, qualified artisans and technicians to aid in successful restoration work, training courses for local technicians in areas such as identification, conservation and rehabilitation of cultural and natural heritage, as well as supplying specialist machinery that the country involved may need but not have the resources to purchase. Subsidies are only considered in the most exceptional of cases.

The major financing of projects does not correspond, in principle, to the international community, and it should be the member state in question that finances each programme and project, except where there resources do not so permit.

To administer international funding of projects, either partially or fully in those cases where it is absolutely necessary, the Paris Convention establishes a trust fund for the protection of the World Cultural and Natural Heritage of Outstanding Universal Value, called "the World Heritage Fund". The fund is replenished through compulsory and voluntary contributions made by member states, by gifts or bequests which may be made by other states or UNESCO, by public or private donations, and by interest accrued on the fund. No political conditions may be attached to contributions to the fund.[31] Also, states undertake to pay regularly, every two years, to the World Heritage Fund, contributions, the amount of which, in the form of a uniform percentage applicable to all States, shall be determined by the General Assembly of States Parties to the Convention.[32]

30 The principle of subsidiarity ensures that an organization will only intervene where the country involved does not have the means to intervene on its own. For a full definition of subsidiarity in a European context, see HAUGHWOUT FOLSOM, R., LAKE, R. & NANDA, V. European Union Law after Maastricht: A Practical Guide for Lawyers Outside the Common Market, Kluwer Law International, 1996, p. 4.

31 Convention of Paris article 15.

32 Convention of Paris article 16.

The convention also obliges the Committee to keep and publish an up to date list of endangered world heritage, under the title of "list of World Heritage in Danger", and for which assistance has been requested under this Convention. This list should contain an estimate of the cost of such operations and should include only such property forming part of the cultural and natural heritage as is threatened by serious and specific dangers, such as:

- the threat of disappearance caused by accelerated deterioration, large-scale public or private projects or rapid urban or tourist development projects;
- destruction caused by changes in the use or ownership of the land;
- major alterations due to unknown causes;
- abandonment for any reason whatsoever;
- the outbreak or the threat of an armed conflict;
- calamities and cataclysms;
- serious fires, earthquakes, landslides;
- volcanic eruptions; and finally
- changes in water level, floods and tidal waves.[33]

At the time of writing 830 World Heritage Sites exist, located in 138 countries. Of these, 644 are cultural, 162 are natural and 24 are mixed properties, and over 40 can be considered as industrial heritage. The endangered list includes 31 properties, including the Dresden–Elbe valley, due to the construction of the "Waldschlösschen Bridge" which would irreversibly damage the values and integrity of the property.[34]

Kiss points out that the importance of the Paris Convention resides in the institutionalization of the principle that, some types of heritage, while under the control of sovereign states, have qualities and values that concern all humanity, and that they should therefore, be conserved by all citizens of the world.[35] Francioni, for his part, argues that one of the weaknesses of the convention is the rigid requisite that the member state should consent to the inscription of a site or an item on the World Heritage List, which could be incompatible with an effective protection

33 Convention of Paris Article 11.4.

34 UNESCO Document WHC-06/30.COM/7B.

35 KISS, A. La notion de Patrimoine commun de l'humanité, Rec de cours de l'Academie de Droit International de La Haye II, 1982, pp. 175 ss.

of world culture. The author argues that there may be many cases where recognition of an item of heritage that represents the interests of minorities may be seen with hostility by the member state. He then goes on to argue that article 4 of the convention, which establishes that the identification of heritage is primarily the responsibility of the member state, does establish, implicitly a possible role for the commission in those cases where the member state will not recognise a particular item of heritage which is clearly representative of human culture, although the nature of the commission makes this clearly impractical.

The consent of the member state is equally a requisite for the concession of funds. The commission cannot provide assistance except when requested to do so by the member state. This is extensible to the list of endangered heritage sites, where the negative of a member state to consent to addition to the list does, in practice, although maybe not in principle, impede the addition to the list.[36]

Nevertheless, as Gemma Prieto points out,

> the relationship between International Law and cultural heritage has to be evaluated cautiously and realistically. While the existence of solemn texts and agreements is undoubtedly important, revealing at least some attempt to reach a consensus view of heritage, there are multiple holes and imperfections that can only be solved by the patient and diligent action of the international organizations called upon to apply these regulations.[37]

3 Industrial Heritage on the World Heritage List

3.1 Why Industrial Heritage is Important

Our industrial heritage is of its very essence the most fragile of commodities. As we advance, we cannot protect all elements of our industrial past. While, in some way or another, everything is relevant to our understanding of our recent history, progress forces us to decide, as a society, what we should try and collectively remember and what we should allow to be forgotten. However, despite, or perhaps, because of, its fragility, the conservation of our industrial heritage is an essential element in our understanding of our recent history. Our industrial

36 FRANCIONI, F. "Treinta años después ¿esta la Convención sobre el Patrimonio Mundial preparada para el siglo XXI" op. cit. p. 37.

37 GEMMA PRIETO, "*Derecho Internacional y bienes culturales*", Revista de Patrimonio Cultural y Derecho, number 2, 1998, p. 20.

heritage is testimony to a whole epoch essential in the formation of our present society.

Table 1: World Heritage Sites in the United Kingdom, the United States and Spain.

UNITED KINGDOM	**UNITED STATES**	**SPAIN**
Blaenavon Industrial Landscape (2000)	*La Fortaleza and San Juan National Historic Site in Puerto Rico (1983)*	*Alhambra, Generalife and Albayzín, Granada (1984, 1994)*
Blenheim Palace (1987)	*Cahokia Mounds State Historic Site (1982)*	*Altamira Cave (1985)*
Canterbury Cathedral, St Augustine's Abbey, and St Martin's Church (1988)	*Chaco Culture (1987)*	*Aranjuez Cultural Landscape (2001)*
Castles and Town Walls of King Edward in Gwynedd (1986)	*Independence Hall (1979)*	*Archaeological Ensemble of Mérida (1993)*
City of Bath (1987)	*Mesa Verde National Park (1978)*	*Archaeological Ensemble of Tárraco (2000)*
Cornwall and West Devon Mining Landscape (2006)	*Monticello and the University of Virginia in Charlottesville (1987)*	*Archaeological Site of Atapuerca (2000)*
Derwent Valley Mills (2001)	*Pueblo de Taos (1992)*	*Burgos Cathedral (1984)*
Durham Castle and Cathedral (1986)	*Statue of Liberty (1984)*	*Catalan Romanesque Churches of the Vall de Boí (2000)*
Frontiers of the Roman Empire (1987, 2005)		*Cathedral, Alcázar and Archivo de Indias in Seville (1987)*
Heart of Neolithic Orkney (1999)		*Historic Centre of Cordoba (1984, 1994)*
Historic Town of St George and Related Fortifications, Bermuda (2000)		*Historic City of Toledo (1986)*
Ironbridge Gorge (1986)		*Historic Walled Town of Cuenca (1996)*
Liverpool – Maritime Mercantile City (2004)		*La Lonja de la Seda de Valencia (1996)*
Maritime Greenwich (1997)		*Las Médulas (1997)*
New Lanark (2001)		*Monastery and Site of the Escurial, Madrid (1984)*
Old and New Towns of Edinburgh (1995)		*Monuments of Oviedo and the Kingdom of the Asturias (1985, 1998)*
Royal Botanic Gardens, Kew (2003)		*Mudejar Architecture of Aragon (1986, 2001)*
Saltaire (2001)		*Old City of Salamanca (1988)*
Stonehenge, Avebury and Associated Sites (1986)		*Old Town of Ávila with its Extra-Muros Churches (1985, 2007)*
Studley Royal Park including the Ruins of Fountains Abbey (1986)		*Old Town of Cáceres (1986)*
Tower of London (1988)		*Old Town of Segovia and its Aqueduct (1985)*
Westminster Palace, Westminster Abbey and Saint Margaret's Church (1987)		*Palau de la Música Catalana and Hospital de Sant Pau, Barcelona (1997)*
		Palmeral of Elche (2000)
		Poblet Monastery (1991)
		Pyrénées - Mont Perdu (1997, 1999)
		Renaissance Monumental Ensembles of Úbed and Baeza (2003)
		Rock Art of the Mediterranean Basin on the Iberian Peninsula (1998)
		Roman Walls of Lugo (2000)
		Route of Santiago de Compostela (1993)
		Royal Monastery of Santa María de Guadalupe (1993)
		San Cristóbal de La Laguna (1999)
		San Millán Yuso and Suso Monasteries (1997)
		Santiago de Compostela (Old Town) (1985)
		University and Historic Precinct of Alcalá de Henares (1998)
		Vizcaya Bridge (2006)
		Works of Antoni Gaudí (1984)

Alfrey and Putnam point out, however, that many see the concept of "industrial heritage" as contradictory, even nonsensical:

> many assumptions about heritage, and the institutions based on them, were formed in a period when an awesome industrialisation transformed familiar landscapes, disrupted habits and challenged established values. Most of what is today protected or celebrated as heritage has been chosen within industrial societies as pre-industrial or non-industrial, as older, more rare, more beautiful, more traditional, more natural, more spiritual. Industrial culture has itself idealised archetypes of what it is not, which can get in the way of recognising its own character, problems and achievements.[38]

Any study of industrial heritage must briefly touch on the process of industrialization, which owes its importance to its global impact on society that produced the largest revolution in human society since the Neolithic period, where humans began to change from hunter–gatherers to an agriculturally based society. Industrialization changed society just as profoundly, transforming its very essence, not only by the metamorphosis of agricultural into industrial society, but also by the very nature and pace of change. The world people leave when they die is now very much more different to the world they were born in to than at any previous time in human history.

In the industrial revolution a new type of building was created, the modern factory. Cities and towns grew under a new concept of industrial urban design, which created what we now call our industrial landscapes. As generations passed, the process of change accelerated, and in the second half of the 20th century, especially from the decade of 1960 onwards, a new type of revolution starts to take place. Commonly known as the post-industrial era, the level of change is at least comparable to the magnitude of change in Neolithic societies or during the first industrial revolution.

In a short number of years society has changed radically – new tools, new ways of working, new forms of media and expression – leaving obsolete, in an amazingly short space of time, machinery and means of production.

The risk of the total destruction and loss of those elements that bear witness to our recent history leads to a growing interest in their preservation, and in the middle of the sixties the first studies of industrial heritage start to appear. One of the first cause celebres of the new discipline was the failed attempt to avoid the destruction of Euston

38 ALFREY, J., PUTNAM, T., The Industrial Heritage, Managing Resources and Uses. Routledge 1992, p. 2.

Station in 1962, converted into a symbol of disregard for the symbols of our recent past. The same period saw the rise of industrial archaeology, a new discipline that has a close relationship with the movement to revalue industrial heritage. Industrial archaeology centred at first on the preservation of buildings, but soon extended to the protection of mobile elements contained in the factories themselves.

Kenneth Hudson, regarded as the first promoter and communicator of industrial archaeology in Europe, defined it as the discovery, cataloguing and study of the physical remains of our industrial past, in order to know through them significant aspects of conditions of work, the technology and the process of production. The object of study would be a wider and deeper knowledge of our industrial past on many different levels, and to know as much as possible, not only about work conditions, but also about worker's and owner's attitudes toward work. To know how and why innovations were made in work processes and by whom. To understand how work was really done – what techniques were used and what type of machinery. To rebuild the physical context in which work was done and where workers, owners and their families lived, and through that, to understand their motivations and desires. Industrial archaeology tries, no less, to understand non-material elements of human society through the reconstruction of the workplace.[39]

As previously seen, understanding industrial historical sites as heritage worthy of protection, although a recent phenomenon is a growing one, and national governments are reflecting these trends by nominating in increasing number industrial sites to the World Heritage List. Some of the more relevant sites are considered below.

3.2 The Zollverein Coal Mine Industrial Complex

The Zollverein XII Coal Mine Industrial Complex was declared to be a world heritage site in 2001. The German commission, proposing the site considered that the site met criteria i, ii, iii and iv, given that

> when the Zollverein Mine XII was completed in 1932, it was considered to be the most modern and beautiful coal mine in the world, the daily output of which, 12,000 tonnes of hard coal, was four times higher than the normal figure. The same year saw the end of the Bauhaus, the most noble objective of which had been to work towards the 'new building of the future' by fusing craft and art. In the opinion of the founder of the Bauhaus, Walter Gropius, the goal of architecture was to create objects and spaces for the purpose of which a new development of form had to proceed, in particular, from the works of

39 HUDSON, K. Industrial Archeology: an Introduction, Oxford University Press 1964.

engineering and industry. At the Zollverein mine the Bauhaus maxim that form must be oriented towards function is perfectly translated into reality. Zollverein XII was created at the end of a phase of political and economic upheaval and change in Germany, which was represented aesthetically in the transition from Expressionism to Cubism and Functionalism. At the same time, Zollverein XII embodies this short economic boom between the two World Wars, which has gone down in history as the 'Roaring Twenties.' However, Zollverein is also, and by no means least, a monument of industrial history reflecting an area in which, for the first time, globalization and the worldwide interdependence of economic factors played a vital part.

The ICOMOS/TICCH committee, after proposing a series of modifications in the candidature, especially referring to the boundaries of the property,[40] felt that the site fulfilled criterion ii (it is considered an exceptional industrial monument by virtue of the fact that its buildings are outstanding examples of the application of the design concepts of the Modern Movement in architecture in a wholly industrial context) and criterion iii (the technological and other structures of Zollverein XII represent a crucial period in the development of traditional heavy industries in Europe, when sympathetic and positive use was made of architectural designs of outstanding quality).[41]

3.3 The Völklingen Ironworks

The Völklingen ironworks in Stadtverband Saarbruken, Saarland, declared world heritage site in 1994, is the only integrated ironworks of the 19th and 20th centuries that went out of production in the recent past to be preserved intact in Western Europe or North America.

According to the proposal by the German UNESCO commission, the site was deemed to have met criterion i, ii and iv. The commission expressed the conviction that

---

40 "The original nomination was of an area based on the previous extent of underground coal-mining concessions at the Zollverein mine. This bears no relationship to surface features, frequently intersecting streets or districts, and does not include all the historic settlements, of which there are at least nine on the map provided with the original nomination. ICOMOS proposed that the nominated area be confined to the Zollverein XII and Zollverein I and II mines, with the adjoining coking plant. This would provide a clear site boundary, bordered by suitable roads and for the most part enclosed by a high wall. It was accepted that the surrounding area is a cultural landscape with many important workers' housing complexes, villas, public buildings, churches, etc, but these would be better treated as the buffer zone to the main industrial complex. The zone should be extended to include the suburbs of Katernberg, Schonnebeck, and Stoppenberg. Consideration should also be given to the inclusion of other important mining sites in the area as part of the nominated area".

41 ICOMOS evaluation document of candidature of the Cultural Industrial Landscape of the Zollverein Mine WHC 4th November 1999.

Dawn Weltkulturerbe Völklinger Hütte by Gerhard Kassner ©Weltkulturerbe Völklinger Hütte/ Gerhard Kassner.

the Völklingen ironworks is a unique monument to the technological history and industrial culture of the 19th and early 20th centuries. It provides an unusually complete illustration of a large pig-iron production plant of major historic interest. The works is a synonym for and a symbol of human achievement during the First and Second Industrial Revolutions; it is a 'cathedral' of the industrial age. The Völklingen ironworks is a unique monument to pig-iron production in Western Europe. As far as is known, no other historic blast-furnace complex has survived which demonstrates the entire process of pig-iron production in the same way, with the same degree of authenticity and completeness and underlined by such a series of technological milestones in innovative engineering. The Völklingen monument illustrates the industrial history of the 19th century in general and of the transnational Saar–Lorraine–Luxembourg industrial region in the heart of Europe in particular.

According to the German commission, proposing the site,

the plant that has survived in Völklingen embodies a world of industrial work and production which has disappeared elsewhere in the wake of subsequent technological development and continuous modernization. The works is to be preserved as an industrial museum and developed as a research facility for iron and steel and for environmental problems. The planned museum park will be able to demonstrate all the stages in the pig-iron production of a former age. All stages of iron production – ore bunker and coking plant (raw-materials preparation and storage) and suspended conveyor system, blast-furnace complex, dry-gas purification plant, and blowing-engine hall – can be experienced directly. In the long term the facilities of the sinter plant and the charge bunker will provide accommodation for the establishment of a European iron and steel museum. The outstanding universal value of the

> Völklingen ironworks lies in its unique completeness and originality. Technological milestones such as the dry gas purification plant, which was the first of its kind on such a large scale, the suspended conveyor system (the largest of its type), and the pioneer sinter plant are all integral parts of a complex 19th and 20th century pig-iron production works concentrated in a small area.[42]

After the evaluation report by ICOMOS the site was deemed to have met criterion ii and iv: criterion ii as several important technological innovations in the production of pig iron were developed or first applied successfully on an industrial scale at Völklingen ironworks and are now in universal use throughout the world, and criterion iv as the Völklingen ironworks is an outstanding example of an integrated pig-iron production plant of the type which dominated this industry in the 19th and early 20th centuries.[43]

Blower Hall Interiour Weltkulturerbe Völklinger Hütte by Franz Mörscher ©Weltkulturerbe Völklinger Hütte/ Franz Mörscher.

42 ICOMOS evaluation document of candidature of the Völklingen ironworks, WHC 15th September 1993.

43 ICOMOS evaluation document of candidature of the Völklingen ironworks, WHC 15th September 1993.

3.4 Saltaire

The United Kingdom has six properties included in the World Heritage List that can be considered industrial heritage. Ironbridge Gorge, Blaenavon Industrial Landscape, New Lanark, Saltaire, Derwent Valley Mills, Cornwall and West Devon mining landscape and the Liverpool Mercantile city.

The case of Saltaire is particularly interesting. According to the evaluation committee Saltaire is a complete and well preserved industrial village of the second half of the 19th century. Its textile mills, public buildings, and workers' housing are built in a harmonious style of high architectural quality and the plan survives intact, giving a vivid impression of the philanthropic paternalism of the Victorian age.[44]

Its significance resides in that the industrial village of Saltaire is an outstanding example of mid 19th century philanthropic paternalism which had a profound influence on developments in industrial and urban planning in the United Kingdom and beyond. It survives in a complete and well preserved form as testimony to the pride and power of basic industries such as textiles for the economy of Great Britain and the world in the 19th and earlier 20th centuries.

ICOMOS recommended that Saltaire be inscribed on the World Heritage List on the basis of criterion ii (Saltaire is an outstanding and well preserved example of a mid 19th century industrial town, the layout of which was to exert a major influence on the development of the "garden city" movement) and criterion iv (the layout and architecture of Saltaire admirably reflect mid 19th century philanthropic paternalism, as well as the important role played by the textile industry in economic and social development).

3.5 Derwent Valley Mills

Another interesting example of industrial heritage is Derwent Valley Mills. According to the ICOMOS document

> the Derwent valley in central England contains a series of 18th and 19th century cotton mills and an industrial landscape of high historic and technological significance. The modern factory owes its origins to the mills at Cromford, where Richard Arkwright's inventions were first put into industrial-scale production. The workers' housing associated with this and the other mills are intact and illustrate the socio-economic development of the area.

---

44 Saltaire was described by William Fairbairn in the work "Treatise on Mills and Millwork" in 1871.

According to ICOMOS/TICCH

> the cultural landscape of the Derwent valley is of outstanding significance because it was here that the modern factory system was established, to accommodate the new technology for spinning cotton developed by Richard Arkwright. The insertion of industrial establishments into a rural landscape necessitated the construction of housing for the workers in the mills, and the resulting settlements created an exceptional industrial landscape that has retained its qualities over two centuries.

The ICOMOS/TICCH committee recommended that the property be inscribed on the World Heritage List on the basis of criterion ii (the Derwent Valley saw the birth of the factory system, when new types of building were erected to house the new technology for spinning cotton developed by Richard Arkwright in the early 19th century) and criterion iv (in the Derwent Valley for the first time there was large-scale industrial production in a hitherto rural landscape. The need to provide housing and other facilities for workers and managers resulted in the creation of the first modern industrial towns).[45]

3.6 Vizcaya Bridge

Spain has three sites that can be considered to include elements of industrial heritage. The first two sites are pre-industrial revolution: the old town of Segovia and its aqueduct, and the Industrial landscape of Las Medulas, monuments or landscapes dating from Roman times.

The latest addition to Spain's world heritage list in 2006[46] is however a post industrial monument, the Vizcaya Bridge. Opened in 1893 it was the first bridge in the world to transport passengers in a hanging transporter. The bridge spans the mouth of the River Ibaizabal and joins two towns, Getxo and Portugalete, west of Bilbao. Its nomination was accepted by the ICOMOS/TICCH commission having fulfilled criterion i in that the bridge was "considered dramatic and aesthetically pleasing addition to the river estuary, and an exceptional expression of technical creativity, reflecting an entirely satisfactory relationship between form and function", and criterion ii because Vizcaya Bridge, "through the development of the hanging transporter mechanism and its fusion of iron working technology with new steel cables, created a new form of construction that influenced the development of bridges around the world over the next three decades".[47]

45 ICOMOS recommendation document Derwent Valley Mills 26 June 2000

46 Decision 30 Com 8B.49.

47 ICOMOS report April 2006.

Boar's Head Mill in Darley Abbey. Courtesy of the Derwent Valley Mills Partnership.

The Vizcaya Bride © El Transbordador De Vizcaya, S.L.

Masson Mills in Matlock Bath. Courtesy of the Derwent Valley Mills Partnership.

Chapter 2

The Protection of Historic Properties in United States Law

The protection of historic properties in the United States is a complex mix of federal, state and local law. The legal framework consists of major statutes dedicated to the protection of historic properties and references to them contained in other statutes, such as those that regulate transport and environmental issues, zoning statutes, agency regulations, tax incentives, and Executive Orders that guide environmental policy. While the federal system provides a framework for protecting significant heritage sites, it is complemented and reinforced at the state level. Indeed, the legal framework established at state level is generally more comprehensive than the federal system. Therefore, in this chapter we will lay out both the relevant Federal legislation and the legislative framework for one of the states with the most significant bodies of cultural heritage regulation: the state of California. We will pay special attention to local ordinances regulating historic property protection – perhaps the most effective tool in the United States legal system for protecting historic buildings.

There are significant differences between the federal and state regulation of historic properties and the related European legal systems. The most outstanding distinction is the non-obligatory nature of the protection of historical buildings in federal and state law. In general, if the property is in private ownership, and if there is no federal funding involved or there is no easement condition attached to the property, then the owner is free to dispose of the property as he or she thinks fit.

On the other hand there is a complete system of funding and incentives at the federal and state level for the preservation of historical heritage, and an approach much more akin to a cooperative effort involving private property owners, trusts, the general public and administrative agencies that tend towards an effective preservation of cultural and historic heritage.

1 Federal Law

The main statute for the protection of historic properties in the United States is the National Historic Preservation Act.[48] Other statutes that are relevant are the National Environmental Policy Act of 1969 (NEPA),[49] the Archaeological and Historic Preservation Act of 1974 (AHPA),[50] the Archaeological Resources Protection Act of 1979 (ARPA),[51] the American Indian Religious Freedom Act of 1978 (AIRFA),[52] the Native American Graves Protection and Repatriation Act of 1990 (NAGPRA)[53] and the Americans with Disabilities Act of 1990 (ADA).[54] There are also relevant elements of protection in the so

48 An Act to establish a programme for the preservation of Additional Historic Properties throughout the Nation, and for other purposes, Approved October 15, 1966 (last amended 1992) Public Law 102-575 16 USC 470.

49 The National Environmental Policy Act of 1969, as amended (Pub. L. 91-190, 42 U.S.C. 4321–4347, January 1, 1970, as amended by Pub. L. 94–52, July 3, 1975, Pub. L. 94–83, August 9, 1975, and Pub. L. 97–258, § 4(b), Sept. 13, 1982.

50 Public Law 93–291 and 16 U.S.C.469–469c – also known as the Archaeological Recovery Act and the Moss–Bennett bill, the act aims to provide for the preservation of historical and archaeological data (including relics and specimens) which might otherwise be irreparably lost or destroyed as the result of (1) flooding, the building of access roads, the erection of workmen's communities, the relocation of railroads and highways, and other alterations of the terrain caused by the construction of a dam by any agency of the United States, or by any private person or corporation holding a license issued by any such agency or (2) any alteration of the terrain caused as a result of any Federal construction project or federally licensed activity or programme.

51 (Public Law 96–95; 16 U.S.C. 470aa–mm) amended four times. The Act was passed with the purpose of securing "for the present and future benefit of the American people, the protection of archaeological resources and sites which are on public lands and Indian lands, and to foster increased cooperation and exchange of information between governmental authorities, the professional archaeological community, and private individuals" (Sec. 2(4)(b)).

52 Public Law 95–341, 42 U.S.C. 1996 and 1996a) amended once.

53 Public Law 101–601–Nov. 16, 1990.

54 The portion of the Act relevant to Historic Properties is Sec. 12204.(c) which lays out that "with respect to alterations of buildings or facilities that are eligible for listing in the National Register of Historic Places under the National Historic Preservation Act (16 U.S.C. 470 et seq.), the guidelines described in paragraph (1) shall, at a minimum, maintain the procedures and requirements established in 4.1.7(1) and (2) of the Uniform Federal Accessibility Standards", and that "with respect to alterations of buildings or facilities designated as historic under State or local law, the guidelines described in paragraph (1) shall establish procedures equivalent to those established by 4.1.7(1)(b) and (c) of the Uniform Federal

called "agency specific legislation" – laws that specifically apply to the actions of administrative agencies. Examples include specific sections of the Department of Transportation Act,[55] the National Forest Management Act of 1976 (NFMA),[56] federal Land Policy and Management Act of 1977 (FLPMA)[57] and the Public Buildings Cooperative Use Act of 1976 (PBCUA).[58]

1.1 Origins of Historic Preservation in United States Law

The first law to address historic property protection at a federal level was the Antiquities Act of 1906.[59] Designed to protect only very small tracts of land around archaeological sites, the Act states that:

> the President of the United States is authorized, in his discretion, to declare by public proclamation historic landmarks, historic and prehistoric structures, and other objects of historic or scientific interest that are situated upon the lands owned or controlled by the Government of the United States to be national monuments, and may reserve as a part thereof parcels of land, the limits of which in all cases shall be confined to the smallest area compatible with the proper care and management of the objects to be protected. When such objects are situated upon a tract covered by a bona fide unperfected claim or held in private ownership, the tract, or so much thereof as may be necessary for the proper care and management of the object, may be relinquished to the Government, and the Secretary of the Interior is authorized to accept the relinquishment of such tracts in behalf of the Government of the United States.

This broad language meant that the Antiquities Act was invoked on many occasions to declare National Monuments. As Squillace[60] points out "at the time it was enacted in 1906, Congress had enacted no other laws that specifically authorized the President to set aside lands for preservation purposes" and that:

Accessibility Standards, and shall require, at a minimum, compliance with the requirements established in 4.1.7(2) of such standards".

55 49 U.S.C. 303 (formerly 49 U.S.C. 1651(b)(2) and 49 U.S.C. 1653f).

56 16 U.S.C. 1600.

57 Public Law 94–57943 U.S.C. 1701, establishes amongst its purposes that "the public lands be managed in a manner that will protect the quality of scientific, scenic, historical, ecological, environmental, air and atmospheric, water resource, and archaeological values.

58 Public Buildings Cooperative Use Act of 1976 (40 U.S.C. 611). The Act encourages adaptive use of historic buildings as administrative facilities for Federal agencies and activities.

59 American Antiquities Act of 1906 16 USC 431–433. For an overview of the origins of Historic Preservation in the United States, see RAND, K "*Nothing lasts forever: toward a coherent theory in American preservation law*" in 27 U. Mich. J.L. Reform 277 1993 pp. 284 and following.

60 SQUILLACE, M. "*The Monumental legacy of the antiquities act of 1906*" in 37 Georgia Law Review 473 Winter 2003.

> between 1906 and 2001, fourteen presidents have established 122 national monuments covering approximately seventy million acres of land in twenty-eight states, one territory, and the District of Columbia, pursuant to their authority under the Antiquities Act. Many of our most treasured national parks – including Grand Canyon, Olympic, Zion, Bryce Canyon, Capitol Reef, Canyonlands, and Glacier Bay – began as national monuments, and were frequently expanded under the Antiquities Act.[61]

The Historic Sites Act of 1935[62] went further, making it a public policy aim to preserve historic sites, buildings and objects of national significance for public use. For Phelan,[63] the important elements of the Historic Sites Act are the declaration that the preservation for public use of historic sites, buildings, and objects of national significance for the inspiration and benefit of the people of the United States should be national policy. To this end, the Secretary of the Interior was authorized to restore, reconstruct, and maintain historic sites and properties and to establish and maintain museums for these purposes. The Secretary was also authorized to contract and enter into cooperative agreements with the states, municipal subdivisions, and private organizations and individuals, in order to protect, preserve, maintain, or operate any historic or archaeological building, site, object, or property connected with a public use.

Neither of these acts, however, attempted to create a broad, inclusive approach to historical heritage preservation, and it wasn't until 1966 with the passage of the National Historic Preservation Act that the legislature at a federal level created a system that allowed the federal government to take a lead, in partnership with state and local governments, in the preservation of historical buildings and sites.

1.2 The National Historic Preservation Act of 1966

Congress, in the passage of the Law, found that although "the spirit and direction of the Nation are founded upon and reflected in its historic heritage", and that "the historical and cultural foundations of the Nation should be preserved as a living part of our community life and development in order to give a sense of orientation to the American people", historic properties significant to the nation's heritage are

61 See LEE, R. The Antiquities Act of 1906. National Park Service, National Technical Information Service, PB88–200837 (1970).

62 49 Stat. 666; 16 U.S.C. 461–467 – amended eight times.

63 PHELAN, M. "*A Synopsis of the Laws Protecting Our Cultural Heritage*" in 28 New England Law. Review 63 Fall 1993 p. 63 y ss.

being lost or substantially altered, "sometimes inadvertently", with increasing frequency.[64]

Congress declared that the preservation of cultural heritage is vital and in the public interest, and that the very same heritage is under severe threat by increased development and urban expansion. The legislature recognised that previous legislative efforts had been ineffective, with the major burdens being borne by private individuals and agencies. The new approach would ensure that these actors would continue to play a vital role, but that it was "nevertheless necessary and appropriate for the Federal Government to accelerate its historic preservation programmes and activities, to give maximum encouragement to agencies and individuals undertaking preservation by private means, and to assist State and local governments and the National Trust for Historic Preservation in the United States to expand and accelerate their historic preservation programmes and activities".[65]

The law lays out that the policy of the federal government should be:

- to foster conditions that allow modern life and historic preservation to coexist;
- to provide leadership in the preservation of the prehistoric and historic resources of the United States;
- to administer federally owned, administered, or controlled prehistoric and historic resources in a spirit of stewardship for the inspiration and benefit of present and future generations;
- to contribute to the preservation of non federally owned prehistoric and historic resources and give maximum encouragement to organizations and individuals undertaking preservation by private means;
- to encourage the public and private preservation and utilization of all usable elements of the Nation's historic built environment; and
- to assist other organizations, amongst whom are state and local governments, to "expand and accelerate their historic preservation programmes and activities".[66]

64 NHPA Section 1 (b).
65 NHPA Section 1 (b) 7.
66 NHPA Section 2.

1.2.1 The National Register of Historic Places

The National Register, created in section 101 of the law, is a central pillar of the National Historic Preservation Act, given that inclusion in this register is a precondition for the receipt of federal funds for the preservation of historic properties. Some authors[67] have pointed out that the National Register does not prevent, in the form of regulation, demolition or alteration by private owners, but rather serves to protect historic and cultural resources by increasing public awareness.

It is an inventory that is composed of districts, sites, buildings, structures and objects significant in American history, architecture, archaeology, engineering and culture. The Secretary of the Interior is the person responsible for listing properties in the National Register, using published criteria (this responsibility has been delegated to the National Parks Service). A property can only be included on the National Register if the owner of that property is in agreement, or does not specifically object to inclusion. In the case that "the owner or owners of any privately owned property, or a majority of the owners of such properties within the district in the case of an historic district, object to such inclusion or designation, such property shall not be included on the National Register or designated as a National Historic Landmark until such objection is withdrawn".[68] In this case, while the Secretary of the Interior does not have the power to force a property's inclusion on the register, he or she does have the right to designate the property as "eligible for inclusion", being obliged thereafter to inform the Advisory Council on Historic Preservation, the appropriate State Historic Preservation Officer, the appropriate chief elected local official and the owner or owners of the property of the decision.[69]

Every four years the Secretary must review, in consultation with the appropriate advisory bodies, properties both on the National Register and eligible for inclusion to determine whether they are under threat of damage or deterioration, determining the kinds of properties that may be threatened, the causes of the threats and developing and submitting to the President and Congress recommendations for appropriate action.[70]

---

67 See for example MCLAMB, S. Preservation law survey 2001: State preservation law in 8 Wid. L. Symposium J. 463.

68 NHPA Section 101(a) 6.

69 NHPA Section 101 (a) 6.

70 NHPA SECTION 101 (a) 7.

The criteria for inclusion of properties on the National Register are included in the Federal Code of Regulations.[71] They are deliberately worded to include a wide diversity of properties.

The evaluation should be based on

> the quality of significance in American history, architecture, archaeology, engineering, and culture present in districts, sites, buildings, structures, and objects that possess integrity of location, design, setting, materials, workmanship, feeling, and association and that are associated with events that have made a significant contribution to the broad patterns of our history; or that are associated with the lives of persons significant in our past; or that embody the distinctive characteristics of a type, period, or method of construction, or that represent the work of a master, or that possess high artistic values, or that represent a significant and distinguishable entity whose components may lack individual distinction; or that have yielded, or may be likely to yield, information important in prehistory or history.

The rules state that "ordinarily cemeteries, birthplaces, or graves of historical figures, properties owned by religious institutions or used for religious purposes, structures that have been moved from their original locations, reconstructed historic buildings, properties primarily commemorative in nature, and properties that have achieved significance within the past 50 years shall not be considered eligible for the National Register".

An excluded property may be considered if it fulfils one of the following conditions:

- It is a religious property deriving primary significance from architectural or artistic distinction or historical importance;
- It is a building or structure removed from its original location but which is significant primarily for architectural value, or which is the surviving structure most importantly associated with a historic person or event;
- It is a birthplace or grave of a historical figure of outstanding importance if there is no appropriate site or building directly associated with his productive life;
- It is a cemetery which derives its primary significance from graves of persons of transcendent importance, from age, from distinctive design features, or from association with historic events;

71 30 CFR part 60.

- It is a reconstructed building when accurately executed in a suitable environment and presented in a dignified manner as part of a restoration master plan, and when no other building or structure with the same association has survived;
- It is a property primarily commemorative in intent if design, age, tradition, or symbolic value has invested it with its own exceptional significance;
- It is a property achieving significance within the past 50 years if it is of exceptional importance.

1.2.2 State Implementation of the National Historic Preservation Act

Section 101 also authorises the Secretary of the Interior to "promulgate or revise regulations for State Historic Preservation Programmes", ensuring that the programme is consistent with the Act. If the state's programme is determined not to be consistent with the National Historic Preservation Act, the Secretary of the Interior may "disapprove the programme and suspend in whole or in part any contracts or cooperative agreements with the State and the State Historic Preservation Officer under this Act, until the programme is consistent with this Act, unless the Secretary determines that the programme will be made consistent with this Act within a reasonable period of time".

The state programme must include provisions for the designation and appointment by the Governor of a "State Historic Preservation Officer" and for the employment or appointment by an officer of such professionally qualified staff as may be necessary for such purposes. It must also have an adequate and qualified State Historic Preservation Review Board designated by the State Historic Preservation Officer and adequate public participation in the State Historic Preservation Programme, including the process of recommending properties for nomination to the National Register.

The State Historic Preservation Officer must, amongst other duties:

- carry out a comprehensive state-wide survey of historic properties, including them in an inventory, identify and nominate eligible properties to the National Register;
- prepare and implement a comprehensive state-wide historic preservation plan;
- administer the State programme of Federal assistance for historic preservation within the State;

- ensure, in cooperation with the appropriate bodies and administrative agencies, that historic properties are taken into consideration at all levels of planning and development;
- set up information and training programmes; and
- cooperate with local governments in the development of local historic preservation programmes.[72]

The law permits State governments to carry out their responsibilities under this act by contract or cooperative agreement with qualified non-profit organizations or educational institutions.[73] It also provides for certification by the State Historic Preservation Officer of local governments to carry out responsibilities under this act. Certification must ensure that the local government enforces appropriate State or local legislation for the designation and protection of historic properties, has established an adequate and qualified historic preservation review commission by State or local legislation, maintains a system for the survey and inventory of historic properties, and provides for adequate public participation in the local historic preservation programme, including the process of recommending properties for nomination to the National Register.[74]

1.2.3 Tribal Properties in the National Historic Preservation Act

Section 101(d) establishes a programme to assist Indian tribes in preserving their particular historic properties, obliging the Secretary of State to foster communication and cooperation between Indian tribes and State Historic Preservation Officers in the administration of the National Historic Preservation Programme in order to ensure that all types of historic properties and all public interests in such properties are given due consideration. Identification, evaluation, protection, and interpretation of historic properties should be coordinated among Indian tribes, State Historic Preservation Officers, and Federal agencies in historic preservation planning, ensuring that Tribal values are taken into account to the extent feasible. To this end, the Secretary of State may waive or modify requirements of the law to conform to the cultural setting of tribal heritage preservation goals and objectives.

At the request of the Tribal chief, the Tribe may take over the functions of the State preservation officer, always given that they present a tribal historic preservation programme which ensures,

72 NHPA Section 101(b) 3.
73 NHPA Section 101(b) 4.
74 NHPA Section 101(c) 1.

amongst other things, that the tribal preservation programme is fully capable of carrying out the functions specified and that the plan defines the remaining responsibilities of the Secretary and the State Historic Preservation Officer.

1.2.4 The Grant Programme

The National Historic Preservation Act authorises a Grant Programme through the National Trust for Historic Preservation. The grants can be divided into two categories. In the first place, there is a provision of funds to States in order to carry out their responsibilities under the Act. In second place, grants are given as direct aid for the renovation and protection of historic properties. These grants may be made for:

- the preservation of National Historic Landmarks which are threatened with demolition or impairment and for the preservation of historic properties of World Heritage significance;
- demonstration projects which will provide information concerning professional methods and techniques having application to historic properties;
- the training and development of skilled labour in trades and crafts, and in analysis and curation, relating to historic preservation;
- the assistance of persons or small businesses within any historic district included in the National Register to remain within the district.

Grants may only be made to protect historically significant religious buildings if "the purpose of the grant is secular, does not promote religion, and seeks to protect those qualities that are historically significant".

The grant programme is a matching grant-in-aid programme, which means that States and other recipients of grants must match the federal funds. The Grantee must also agree to assume, after completion of the project, the total cost of the continued maintenance, repair, and administration of the property.

In fiscal year 2005, the US Congress appropriated $71.6 million from the HPF for historic preservation grants.[75]

75 Source Historic Preservation Fund 2005 Report.

1.2.5 Advisory Council on Historic Preservation

Section 201 of the National Historic Preservation Act creates the now independent agency, the Advisory Council on Historic Preservation, which is composed of 20 members appointed by the president[76].

The Council has a wide range of functions relating to the preservation of historic properties including:

- advising the President and the Congress on matters relating to historic preservation;
- encouraging, in cooperation with the National Trust for Historic Preservation and appropriate private agencies, public interest and participation in historic preservation;
- recommending the conduct of studies in such areas as the adequacy of legislative and administrative statutes and regulations pertaining to historic preservation activities of State and local governments and the effects of tax policies at all levels of government on historic preservation;
- encouraging, in cooperation with appropriate public and private agencies and institutions, training and education in the field of historic preservation;
- reviewing the policies and programmes of Federal agencies and recommend to such agency's methods to improve the effectiveness, coordination, and consistency of those policies and programmes with the policies and programmes carried out under the National Historic Preservation Act.[77]

The Council is also authorized to promulgate rules which affect what is, in effect, the most significant section of the act, Section 106.[78]

---

76 Section 201 (a) The council is comprised of a Chairman appointed by the President selected from the general public, the Secretary of the Interior, the Architect of the Capitol, the Secretary of Agriculture and the heads of four other agencies of the United States (other than the Department of the Interior), the activities of which affect historic preservation, one Governor appointed by the President, one mayor appointed by the President, the President of the National Conference of State Historic Preservation Officers, the Chairman of the National Trust for Historic Preservation, four experts in the field of historic preservation from the disciplines of architecture, history, archaeology, and other appropriate disciplines, three at-large members from the general public, appointed by the President and one member of an Indian tribe or Native Hawaiian organization.

77 Section 202 (a).

78 Section 211 The Council is authorized to promulgate such rules and regulations as it deems necessary to govern the implementation of section 106 of this Act in its entirety. The Council shall, by regulation, establish such procedures as may be necessary to provide for participation by local governments in proceedings and other actions taken by the Council with respect to undertakings referred to in section 106 which affect such local governments.

These regulations set forth administrative procedures, known as the "Section 106 Process", and lay out how federal agencies must take into account how their actions will affect historic properties.

1.2.6 Section 106 of the National Historic Preservation Act

The part of the National Historic Preservation Act which has been most significant in the protection of historic properties, and which has generated a large body of case law, is section 106, which establishes that:

> The head of any Federal agency having direct or indirect jurisdiction over a proposed Federal or federally assisted undertaking in any State and the head of any Federal department or independent agency having authority to license any undertaking shall, prior to the approval of the expenditure of any Federal funds on the undertaking or prior to the issuance of any license, as the case may be, take into account the effect of the undertaking on any district, site, building, structure, or object that is included in or eligible for inclusion in the National Register. The head of any such Federal agency shall afford the Advisory Council on Historic Preservation established under Title II of this Act a reasonable opportunity to comment with regard to such undertaking.

Therefore, a federal agency must do two things before carrying out or funding a project the may effect properties included, or eligible for conclusion, on the National Historic Register. In the first place, the agency must take into account the effects of their actions on historic properties. In second place, they must consult with the Advisory Council on Historic Preservation about the project.

The administrative procedures that govern this process are contained in 36 CFR PART 800 – protection of Historic Properties. According to the agency, "the Section 106 process seeks to accommodate historic preservation concerns with the needs of Federal undertakings through consultation among the agency official and other parties with an interest in the effects of the undertaking on historic properties, commencing at the early stages of project planning". For the agency, "the goal of consultation is to identify historic properties potentially affected by the undertaking, assess its effects and seek ways to avoid, minimize or mitigate any adverse effects on historic properties".

Compliance with section 106 must take place before the approval of federal funds and before the issuing of a licence for any undertaking covered by this process.[79] This does not however prohibit non-

---

79 Undertaking means a project, activity, or programme funded in whole or in part under the direct or indirect jurisdiction of a Federal agency, including those carried out by or on behalf

destructive planning before 106 compliance is fulfilled. The rules do advise that compliance efforts are begun at an early stage in planning so that a wide range of alternatives may be considered.[80]

In general terms, the steps to be followed are as follows:

- Determine whether the project is the type that has a potential effect on historic properties. If the undertaking is a type of activity that does not have the potential to cause effects on historic properties, assuming such historic properties were present, the agency official has no further obligations under section 106;
- In the case that it does have the potential to affect historic properties, the agency involved should contact the appropriate State Historic Preservation Officer, and begin the process of identifying all historic properties in the area that may be affected. The agency should follow either the Advisory Council on Historic Preservation regulations or alternative agency regulations if they have been approved;
- If there will be no effect on the historic properties identified, then the agency may proceed with the project after consultation with the State Historic Preservation Officer and receiving no objection after 15 days;
- If there is likely to be an adverse effect, a consultation process begins where the agency and the State Historic Preservation Officer try to reach agreement on ways to mitigate the effects on historic properties. If the agency official and the State Historic Preservation Officer agree on how the adverse effects will be resolved, they shall execute a Memorandum of Agreement which is, in effect, a binding contract governing implementation of the project.[81]

---

of a Federal agency; those carried out with Federal financial assistance; and those requiring a Federal permit, license or approval. 30 CFR 800.16(y).

80 30 CFR PART 800 SUBPART A 800.1.

81 Grimm points out the limited effect that section 106 has on damaging Indian Religious sites, stating that while "Traditional Cultural Property identification has helped to avoid unknowing impacts to sacred lands, where Indians can reveal such information, Traditional Cultural Property identification does not, however, prevent an agency from adversely affecting such sites, so long as the agency has followed the correct processes and considered the views of Indians. For that reason, Traditional Cultural Property identification alone is not a particularly effective mechanism to ensure protection of sacred sites, as agencies can, and do, decide to proceed with projects despite impacts to Traditional Cultural Properties". GRIMM, L "Sacred Lands and the Establishment Clause: Indian Religious Practices on

A) Alternative Procedures, Programme Alternatives and Exemptions

Revisions to the rules have created the possibility of Federal Agency Programme alternatives that increase flexibility in the administrative procedure.

In first place, an agency may develop alternate administrative procedures to implement section 106 and substitute them for the Advisory Council on Historic Preservation regulations, always given that they are consistent with the Council's regulations. The development of alternate administrative procedures is subject to a period of consultation, and Advisory Council on Historic Preservation approval. Once approval is given, the agency must notify the parties with which it has consulted and publish notice of final alternate administrative procedures in the Federal Register. These alternate administrative procedures will then substitute the Council regulations insofar as compliance with section 106 of the National Historic Preservation Act is concerned.

In second place, if a long term programme is involved, an agency may negotiate a Programmatic Agreement. These may be developed in the following cases:

- where effects on historic properties are similar and repetitive or are multi-State or regional in scope;
- where effects on historic properties cannot be fully determined prior to approval of an undertaking;
- when non federal parties are delegated major decision making responsibilities;
- where routine management activities are undertaken at Federal installations, facilities, or other land management units; or
- where other circumstances warrant a departure from the normal section 106 process.[82]

The process for approval of a Programmatic Agreement involves consultation with, as appropriate, State Historic Preservation Officers, the National Conference of State Historic Preservation Officers, Indian

Federal Lands", 12 Nat. Resources & Environment 19 (1997) p. 19. Brady underlines this approach, stating that the NHPA is of doubtful efficacy on its own, given its inability to prevent federal agencies that follow proper procedure from affecting sacred sites always given that they follow proper procedure BRADY, J. "*Land is itself a sacred, living being. Native American sacred site protection on federal public lands amidst the shadows of Bear Lodge*", in 24 American Indian Law. Review 153.

82 30 CFR 800.14 (b) 1.

tribes and Native Hawaiian organizations, other Federal agencies, and members of the public. With reference to public participation, the agency official must consider the nature of the programme and its likely effects on historic properties and take steps to involve the individuals, organizations and entities likely to be interested.[83]

If the Advisory Council on Historic Preservation determines that the terms of a programmatic agreement are not being carried out, or if such an agreement is terminated, the agency will have to follow the 106 process for all undertakings included in the programme.[84]

The regulations also establish that an administrative agency may propose a type of action or project that can be exempted from fulfilling the 106 process. The exemption must fulfil the following criteria:

- the actions within the programme or category would otherwise qualify as "undertakings";
- the potential effects of the undertakings within the programme or category upon historic properties are foreseeable and likely to be minimal or not adverse; and
- exemption of the programme or category is consistent with the purposes of the act.

The approval of the exemption is subject to a consultation process, a period of public participation and review by the Advisory Council on Historic Preservation.[85] When an undertaking falls within an approved exemption category, then the 106 process need not be followed, "unless the agency official or the Council determines that there are circumstances under which the normally excluded undertaking should be reviewed".[86]

The rules also contemplate the possibility of the creation of a set of "standard methods" for the treatment of a category of historic properties, a category of undertakings, or a category of effects on historic properties.

B) Identifying Historic Properties Affected by the 106 Process

Two of the most important features of the 106 process are the scope of the search for historic property that is necessary, and the determination of the type of property which triggers the 106 process.

83 30 CFR 800.14 (b) 2.ii.
84 30 CFR 800.14 (b) 2.v.
85 30 CFR 800.14 (c) 1.
86 30 CFR 800.14 (c) 6.

The regulations oblige the agency to carry out a widespread investigation into the historic properties that are within the scope of their undertaking, or within the "area of potential effect". The area of potential effect is defined as "the geographic area or areas within which an undertaking may directly or indirectly cause alterations in the character or use of historic properties, if any such properties exist. The area of potential effects is influenced by the scale and nature of an undertaking and may be different for different kinds of effects caused by the undertaking".[87] The courts have differed in their interpretation of this prevision, in some cases suggesting that the area of potential effects should go beyond the perimeters of the actual project,[88] and in others, the area of potential effects has been determined simply as the boundaries of the project.

In order to identify potential historic properties in the area of potential effects, the agency must conduct a review of existing information on historic properties, including any data concerning possible historic properties not yet identified. The agency must also seek information, as appropriate, from consulting parties, and other individuals and organizations likely to have knowledge of, or concerns with, historic properties in the area, and identify issues relating to the undertaking's potential effects on historic properties. Special administrative procedures are also in place for Indian and Hawaiian properties.

The regulations state that the agency should make a "reasonable and good faith effort to carry out appropriate identification efforts, which may include background research, consultation, oral history interviews, sample field investigations and field surveys".[89] The regulations do concede that the level of effort will depend on the nature and scope of the project and its probable effects on historic properties.

Until 1976, the process was triggered only by properties listed on the National Register of Historic Places. The 1976 reform of the law introduced the provision that properties officially determined to be "eligible" to be included on the National Register triggered the process. Therefore, not only the officially determined properties

---

87 30 CFR 800.16(d).

88 For example, *Colorado River Indian Tribes v. Marsh*, 605 F. Supp. 1425 (C.D. Cal. 1985).

89 30 CFR 800.4 (b) 1.

triggered the process, but also properties that fulfilled the criteria to be eligible, but were not yet declared as such.[90]

After the completion of the investigative phase which has identified "potentially historic properties", the agency has to determine whether they should pass from "potential" to be considered as fully fledged historic properties. In order to do this, the agency should apply National Register Criteria, discussed above. A property that has previously been considered as not eligible may have to be reconsidered in light of "the passage of time, changing perceptions of significance, or incomplete prior evaluations".[91]

If the agency official determines that any of the criteria has been met, and the State Historic Preservation Officer agrees, then the property is considered to be "eligible". If they both agree the criteria have not been met, then the property is considered not eligible.

In the case that the agency official does not believe the criteria have been met, but the State Historic Preservation Officer disagrees, or if the Council or Secretary so request, then a determination of eligibility will have to be made by the Secretary.[92]

There are two possible results of this process: no properties found, and properties adversely affected. In the first of these cases: no properties found, the agency advises all relevant parties and publishes its findings. If no objections are made in 30 days by the State Historic Preservation Officer, then the agency's obligations under section 106 are considered to be fulfilled.

If the agency does object, then a process is opened either of consultation with the objecting party, or of consultation with the Advisory Council on Historic Preservation. The Council's opinion is, however, not binding. The agency can publish a finding contrary to the Council's opinion, but must "prepare a summary of the decision that contains the rationale for the decision and evidence of consideration of the Council's opinion, and provide it to the Council, the State Historic Preservation Officer, and the consulting parties".[93] If the final decision of the agency is to affirm the initial agency finding of no historic properties affected, once the summary of the decision has been notified

90 In the case *Boyd v. Roland*, 789 F.2d 347 (5th Cir. 1986), the court, reviewing the regulations that implement Section 106, found that "eligible property" includes any property that meets National Register criteria, even if it has not been formally determined as meeting the criteria.

91 30 CFR 800.4 (c) 1.

92 30 CFR 800.4 (c) 2.

93 30 CFR 800.4 (d) 1. iv. C.

to all interested parties, the agency's responsibilities have been fulfilled in regards to section 106.

If historic properties are found to have been "possibly adversely affected", then an administrative procedure designed to avoid adverse effects is initiated. In the first place, the type of adverse effects must be identified.[94] The agency may find that there will be no adverse effect, in which case they must again notify the State Historic Preservation Officer and other parties. If there is no objection, then the administrative procedure is closed. If the State Historic Preservation Officer objects, once again a process of review is begun, which may involve requesting the opinion of the Advisory Council on Historic Preservation, which again, is not binding, but which must, however, be taken into account in the agency's published decision.

In the case that adverse effects are found, then there is an obligation to consider alternative projects, or modifications to the existing project that will eliminate or minimise damage to historic properties. After a consultation period, if the agency and the State Historic Preservation Officer agree on how to resolve possible adverse effects on Historical properties, they execute a Memorandum of Agreement. The Memorandum can be either by direct agreement between the two parties, or with Council involvement.[95] In the first case the signatories are the agency and the State Historic Preservation Officer, while in the second case the Council also appears as a signatory. The agency may also invite other parties to become signatories to the agreement, and must invite any party that assumes responsibilities under the

94 30 CFR 800.5 (a) 2. Examples given of adverse effects are:
- Physical destruction of or damage to all or part of the property;
- Alteration of a property, including restoration, rehabilitation, repair, maintenance, stabilization, hazardous material remediation and provision of handicapped access, that is not consistent with the Secretary's Standards for the Treatment of Historic Properties (36 CFR part 68) and applicable guidelines;
- Removal of the property from its historic location;
- Change of the character of the property's use or of physical features within the property's setting that contribute to its historic significance;
- Introduction of visual, atmospheric or audible elements that diminish the integrity of the property's significant historic features;
- Neglect of a property which causes its deterioration, except where such neglect and deterioration are recognized qualities of a property of religious and cultural significance to an Indian tribe or Native Hawaiian organization; and
- Transfer, lease, or sale of property out of Federal ownership or control without adequate and legally enforceable restrictions or conditions to ensure long-term preservation of the property's historic significance.

95 30 CFR 800.6 (b) 1–2.

Memorandum. All signatories have equal rights whenever the Memorandum must be modified.

The Memorandum is evidence of the agency's compliance with section 106. The agency must insure that the project is carried out according to all its provisions. It may include provisions for reporting and monitoring as well as for termination and reconsideration of terms if the project is not terminated on time. It may also contain provisions in the case of future discovery of historic sites, and may be altered if all signatories agree. If any of the parties is not satisfied with the execution of the Memorandum, they may seek to amend or terminate it.

C) Coordinating the 106 Process with the National Environmental Policy Act of 1969

The National Environmental Policy Act[96] was passed in recognition of "the profound impact of man's activity on the interrelations of all components of the natural environment, particularly the profound influences of population growth, high-density urbanization, industrial expansion, resource exploitation, and new and expanding technological advances and recognizing further the critical importance of restoring and maintaining environmental quality to the overall welfare and development of man, declares that it is the continuing policy of the Federal Government, in cooperation with State and local governments, and other concerned public and private organizations, to use all practicable means and measures, including financial and technical assistance, in a manner calculated to foster and promote the general welfare, to create and maintain conditions under which man and nature can exist in productive harmony, and fulfil the social, economic, and other requirements of present and future generations of Americans".[97]

In fulfilment of these objectives, NEPA establishes a process which will insure that "presently unquantified environmental amenities and values may be given appropriate consideration in decision making along with economic and technical considerations". This is done by forcing the responsible agency to prepare a detailed statement listing the environmental impact of the proposed action, any adverse environmental effects which cannot be avoided should the proposal be implemented, alternatives to the proposed action, the relationship

96 The National Environmental Policy Act of 1969, as amended (Pub. L. 91–190, 42 U.S.C. 4321–4347, January 1, 1970, as amended by Pub. L. 94–52, July 3, 1975, Pub. L. 94–83, August 9, 1975, and Pub. L. 97–258, § 4(b), Sept. 13, 1982).

97 Sec. 101 [42 USC § 4331].

between local short-term uses of man's environment and the maintenance and enhancement of long-term productivity, and any irreversible and irretrievable commitments of resources which would be involved in the proposed action should it be implemented.[98]

While early cases understood NEPA to give a substantive mandate to agencies to mitigate environmental damage, later court cases created jurisprudence in another direction – interpreting NEPA as a law that only forces agencies to *consider* project alternatives, allowing them to take a final decision based on agency discretion. NEPA is now considered procedural, not substantive.[99]

Despite its non-substantive nature, agencies do have to fulfil their procedural obligations under NEPA as well as the procedural obligations given to them by the section 106 process. In order to avoid duplication of agency investigations, the section 106 process should be coordinated at the earliest stage with the National Environmental Policy Act. The determination of whether an undertaking is a "major Federal action significantly affecting the quality of the human environment," requiring the elaboration of an environmental impact statement should include a determination of whether the project will have an adverse effect on historic properties. However, a finding of adverse effect on a historic property is not an automatic trigger that requires an Environmental Impact Statement.

In fulfilment of their NEPA obligations, agency officials should ensure that preparation of an environmental assessment and finding of no significant impact or an Environmental Impact Statement and record of decision includes appropriate scoping, identification of historic properties, assessment of effects upon them, and consultation leading to resolution of any adverse effects.[100] The fact that a project may be categorically excluded from NEPA does not mean that the agency does not have to fulfil its section 106 responsibilities.

In order to avoid a duplication of administrative procedures, the rules on 106 compliance allow the agency to incorporate section 106 administrative procedures within those also included in NEPA, The State Historic Preservation Officer must be advised, and certain

---

98 Sec. 102 [42 USC § 4332].2 (c).

99 See in this sense *Weinberger v. Catholic Action of Hawaii*, 454 US 139, 143 (1981) which stated that the need to prepare an EIS is sufficient to ensure that environmental considerations are injected into the federal agency's decision making process and *Kleppe v. Sierra Club*, 427 U.S 390, 410 n.21 (1976) which stated that NEPA ensures that agencies take a "hard look" at the environmental consequences of their actions).

100 300 CFR 800 (a) 3.

standards for developing environmental documents to the 106 process must be met.

D) Special Requirements for National Historic Landmarks

The precautionary nature of Section 106 is strengthened when the historic properties involved are National Historic Landmarks. The National Historic Preservation Act states that "prior to the approval of any Federal undertaking which may directly and adversely affect any National Historic Landmark, the head of the responsible Federal agency shall, to the maximum extent possible, undertake such planning and actions as may be necessary to minimize harm to such landmark, and shall afford the Advisory Council on Historic Preservation a reasonable opportunity to comment on the undertaking".[101]

This legislative imperative means that the agency must request the Council to participate in any consultation to resolve adverse effects on National Historic Landmarks and must notify the Secretary of State of any consultation involving a National Historic Landmark and invite him to participate in the consultation where there may be an adverse effect.

In the case that the Council has participated in consultation, it has the obligation to report the outcome of the section 106 process, including its written comments or any memoranda of agreement to which it is a signatory, to the Secretary of the Interior and the head of the agency involved.

1.3 National Heritage Areas

There is at this moment no specific enabling act relative to all National Heritage Areas, and Congress has created them on an ad hoc basis, each having its own Act specifying its conditions and funding. In order to fully explain the system, reference is made to a bill now before Congress which intends to unify the National Heritage Areas system,[102] describing a National Heritage area as "an area or corridor designated by Congress that tells nationally important stories representing...American heritage". The bill points out that "certain areas of the United States...illustrate significant aspects of our heritage; possess exceptional natural, cultural, scenic, and historic resources; and represent the diversity of our national character" Additionally, in these areas, "the interaction of natural processes, geography, history,

101 NHPA section 110 (f).

102 S. 278: National Heritage Areas Partnership Act.

cultural traditions, and economic and social forces form distinctive landscapes that should be recognized, protected, enhanced, and interpreted to improve the quality of life in the regions and to provide opportunities for public appreciation, education, enjoyment, and economic sustainability".[103]

With the intention of establishing a system of regional and community-based National Heritage Areas to protect, enhance, and interpret natural, historic, scenic, and cultural resources that together tell nationally important stories representing the country's heritage, the Act intends to foster local and community initiatives. These initiatives would be formed by promoting partnerships between Federal, Tribal, State and local governments, non-profit organizations, the private sector and citizens, providing assistance to set "resource stewardship and interpretive goals, and to implement strategies for resource protection and renewed economic viability in these areas".[104] The Act also intends to provide a unified framework for National Heritage Areas, which up until this date had been approved individually by Acts of Congress without a unifying legislative framework. As pointed out by the general accounting office, up until this date, no systematic process exists for identifying qualified sites and designating them as National Heritage Areas. The report continues

> while the Congress generally has designated heritage areas with the Park Service's advice, it designated 10 of the 24 areas without a thorough agency review; in 6 of these 10 cases, the agency recommended deferring action. Even when the agency fully studied sites, it found few that were unsuitable. The agency's criteria are very general. For example, one criterion states that a proposed area should reflect 'traditions, customs, beliefs, and folk life that are a valuable part of the national story.' These criteria are open to interpretation and, using them, the agency has eliminated few sites as prospective heritage areas.[105]

A framework is also provided by the Act to authorize the Secretary of the Interior to provide financial and technical assistance to local coordinating entities that act as a catalyst for diverse regions, communities, organizations, and citizens to undertake projects and programmes for resource stewardship and interpretation.[106]

The Act would establish a National Heritage Areas system which is composed of previous National Heritage Areas and National Heritage

---

103 National Heritage Areas Partnership Act, Section 2 (a) 1–2.

104 National Heritage Areas Partnership Act. Section 2 (a) 5.

105 GAO-04-593T p.

106 National Heritage Areas Partnership Act Section 2 (b) 6.

Areas designated by Congress subsequent to the date of enactment of the Act. National Heritage Areas are not considered to be units of the National Park System and are not subject to the authorities applicable to units of the National Park System.

The Secretary of the Interior would be given responsibilities to conduct feasibility studies for the designation of National Heritage Areas, submitting an annual report to Congress and conducting an evaluation and preparing a report on the accomplishments, sustainability, and recommendations, if any, for the future of each designated National Heritage Area three years prior to the cessation of Federal funding for the area.

The feasibility study examines whether the proposed National Heritage Area:

- has an assemblage of natural, historic, or cultural resources that together tell a nationally important story;
- represents distinctive landscapes and aspects of American heritage worthy of recognition, conservation, interpretation, and continuing use;
- is best managed as such an assemblage through partnerships among public and private entities at the local or regional level;
- reflects traditions, customs, beliefs, and folk life that are a valuable part of the national story;
- provides outstanding opportunities to conserve natural, cultural, historic, and/or scenic features;
- provides outstanding recreational and educational opportunities; and
- has the resources and traditional uses important to the identified stories and themes and these resources and uses retain a degree of integrity capable of interpretation.[107]

The study would also examine whether residents, business interests, non-profit organizations, and governments are involved in the planning and have demonstrated significant support for National Heritage Area designation and whether the local coordinating entity responsible for preparing and implementing the management plan has been identified. The proposed local coordinating entity must have a significant commitment to work in partnership to protect, enhance and

107 Section 5 (a) 1.

interpret resources within the National Heritage Area, having developed a feasible conceptual financial plan that outlines the roles of all participants including the Federal government. A conceptual boundary map should be developed and be supported by the public, as well as by participating Federal agencies. The proposal must also be consistent with continued economic activity within the area.[108]

Once a feasibility study has been conducted and transmitted to congress, the area would be designated by an Act of Congress.[109] Once an area is designated, the local coordinating entity would have to prepare a management plan within three years from the date when any funds are made available for this purpose after designation as a National Heritage Area.[110] The management plans must, amongst others, describe comprehensive policies, goals, strategies, and recommendations for telling the story of the region's heritage and encouraging long-term resource protection, enhancement, interpretation, funding, management, and development of the National Heritage Area.[111] It must also include an inventory of the natural, historical, cultural, educational, scenic, and recreational resources of the National Heritage Area related to the stories and themes of the region that should be protected, enhanced, managed or developed,[112] and include a business plan that provides adequate assurances that the local coordinating entity has the partnerships, and financial and other resources necessary to implement the management plan for the National Heritage Area.[113]

1.3.1 National Heritage Areas and Property Rights

A National Heritage Area is based on a partnership between many different stakeholders, and is a formula that will allow the promotion and preservation of certain features that make the area special. It does not, however, in any way affect any property rights.

In this sense, the Act lays out very specifically that nothing in the Act shall be construed to:

– abridge the rights of any property owner, whether public or private, including the right to refrain from participating in any

108 Section 5 (a) 6.
109 Section 6 (a) 1.
110 Section 7 (b) 1.
111 Section 7 (a) 1.
112 Section 7 (a) 4.
113 Section 7 (a) 8 B.

plan, project, programme, or activity conducted within the National Heritage Area;

- require any property owner to permit public access (including Federal, Tribal, State, or local government access) to such property;
- alter any duly adopted land use regulation or any approved land use plan or any other regulatory authority of any Federal, State, or local agency or Tribal government, or to convey any land use or other regulatory authority to any local coordinating entity;
- authorize or imply the reservation or appropriation of water or water rights;
- diminish the authority of the State to manage fish and wildlife including the regulation of fishing and hunting within the National Heritage Area;
- create any liability, or to have any effect on any liability under any other law, of any private property owner.

Backing this interpretation, the General Accounting Office in 2004, in a report on accountability in National Heritage Areas, concluded that

> Heritage areas do not appear to have affected property owners' rights. In fact, the designating legislation of 13 areas and the management plans of at least 6 provide assurances that such rights will be protected. However, property rights advocates fear the effects of provisions in some management plans. These provisions encourage local governments to implement land use policies that are consistent with the heritage areas' plans, which may allow the heritage areas to indirectly influence zoning and land use planning in ways that could restrict owners' use of their property. Nevertheless, heritage area officials, Park Service headquarters and regional staff, and representatives of national property rights groups that we contacted were unable to provide us with any examples of a heritage area directly affecting – positively or negatively – private property values or use.[114]

In this sense, the bill introduced to establish the South Park National Heritage Area states that "no privately owned property shall be preserved or promoted under the management plan for the Heritage Area until the owner of that private property has been notified in writing by the management entity and given written consent to the management entity for such preservation or promotion". Furthermore, "any owner of private property included within the boundary of the

114 GAO-04-593T.

Heritage Area shall have their property immediately removed from the boundary by submitting a written request to the management entity".[115]

It is worth indicating, however, that opponents of the National Heritage Area system allege that property rights are affected by the influence it may have on zoning plans. Chumley and Utt point out:

> that suggested courses of action, as put forth in the management plans required as part of National Heritage Area designation and subsequent oversight, do in fact compromise the ability of property owners to use land as they see fit. It is one thing to opt out of the National Heritage Area declaration, which is allowed in the accompanying legislation and regulation, but it is another thing entirely to try to opt out of ordinances and laws created by the local government at the recommendation of the National Park Service or the management entity.[116]

National Heritage Areas are being created with increased frequency with 10 of the 37 areas being designated in 2006.[117] Additionally, 15 designation bills are now before Congress.

---

115 South Park National Heritage Act (Introduced in House), 110th Congress, August 2, 2007,

116 CHUMLEY & UTT National Heritage Areas: Costly Economic Development Schemes that Threaten Property Rights, in Backgrounder No. 2080 October 23, 2007. National Heritage Foundation.

117 National Heritage areas and their date of designation are:

- 1984 Illinois & Michigan Canal National Heritage Corridor IL;
- 1986 John H. Chafee Blackstone River Valley National Heritage Corridor MA/RI;
- 1988 Delaware and Lehigh National Heritage Corridor PA;
- 1988 Southwestern Pennsylvania Heritage Preservation Commission PA;
- 1994 Cane River National Heritage Area LA;
- 1994 Quinebaug and Shetucket Rivers Valley National Heritage Corridor CT/MA;
- 1996 America's Agricultural Heritage Partnership IA;
- 1996 Augusta Canal National Heritage Area GA;
- 1996 Cache La Poudre River Corridor CO;
- 1996 Essex National Heritage Area MA;
- 1996 Hudson River Valley National Heritage Area NY;
- 1996 National Coal Heritage Area WV;
- 1996 Ohio & Erie Canal National Heritage Corridor OH;
- 1996 Rivers of Steel National Heritage Area PA;
- 1996 Shenandoah Valley Battlefields National Historic District VA;
- 1996 South Carolina National Heritage Corridor SC;
- 1996 Tennessee Civil War Heritage Area TN;
- 1998 Automobile National Heritage Area MI;
- 2000 Erie Canalway National Heritage Corridor NY;
- 2000 Lackawanna Valley Heritage Area PA;
- 2000 Schuylkill River Valley National Heritage Area PA;
- 2000 Wheeling National Heritage Area WV;
- 2000 Yuma Crossing National Heritage Area AZ;
- 2003 Blue Ridge National Heritage Area NC;
- 2004 Mississippi Gulf National Heritage Area MS;
- 2004 National Aviation Heritage Area OH;

1.3.2 Rivers Of Steel Heritage Area

An example of how the National Heritage Areas scheme has been useful in protecting a wide industrial area is the Rivers of Steel Heritage Area. Finding that "the industrial and cultural heritage of south-western Pennsylvania, including the city of Pittsburgh, and the counties of Allegheny, Armstrong, Beaver, Fayette, Greene, Washington, and Westmoreland, related directly to steel and steel-related industries, is nationally significant", Congress designated in 1996[118] an area known initially as the Steel Industry Heritage Area.

Congress felt that these industries, including steel making, iron making, aluminium, speciality metals, glass, coal mining, coke production, machining and foundries, transportation, and electrical industries had created an industrial and cultural heritage that included the social history and living cultural traditions of the people of the region. Also the labour movement of the region had played a significant role in the development of the Nation, including the formation of many key unions such as the Congress of Industrial Organizations (CIO) and the United Steel Workers of America (USWA), and crucial struggles to improve wages and working conditions, such as the Rail Strike of 1877, the Homestead Strike of 1892, and the Great Steel Strike of 1919.[119]

Steel production in the area began with the construction of the Andrew Carnegie Steel Mill in 1875, and by the early 20th century, the area became heavily industrialised, becoming the base of some of the most important companies in the America, amongst which were US Steel, Westinghouse, Alcoa, H.J. Heinz, J.K. Porter, and Mellon Bank. In the 1970s and 1980s many sectors in the area entered into decline. The Rivers of Steel Heritage area was formed with the object of promoting tourism, protecting and interpreting historic and cultural resources, creating education and interpretative programmes, and

---

- 2004 Oil Region National Heritage Area PA;
- 2006 Arabia Mountain National Heritage Area GA;
- 2006 Atchafalaya National Heritage Area LA;
- 2006 Champlain Valley National Heritage Partnership NY/VT;
- 2006 Crossroads of the American Revolution National Heritage Area NJ;
- 2006 Freedom's Frontier National Heritage Area KS/MO;
- 2006 Great Basin National Heritage Route UT/NV;
- 2006 Gullah/Geechee Cultural Heritage Corridor NC/SC;
- 2006 Mormon Pioneer National Heritage Area UT;
- 2006 Northern Rio Grande National Heritage Area NM;
- 2006 Upper Housatonic Valley National Heritage Area MA/CT .

118 Public Law 104–333, section 401.

119 Public Law 104–333, section 402 (a).

revitalizing river valleys from industrial corridors into a system of green ways linking natural, historical and recreational sites.

In the ten years since its foundation, the heritage area has been involved in more than 250 projects to revitalise the area and to preserve its industrial heritage. The projects have been carried out with a variety of goals: protection of historic resources, maintaining cultural traditions, promoting economic development as well as creating educational resources to foment interest in the historic heritage represented by the area. Since its creation, over 37 million dollars have been spent on projects, with nearly 10 million dollars being destined to historical preservation. Over half these funds came from the National Park Service and other federal funding programmes.[120]

1.3.3 Upper Housatonic Valley National Heritage Area

Another example of how a Heritage Area is used to protect a particular area is the Upper Housatonic Valley National Heritage Area. Declared a National Heritage Area in 2006, it is considered by Congress to encompass "29 towns in the hilly terrain of western Massachusetts and north-western Connecticut, is a singular geographical and cultural region that has made significant national contributions through its literary, artistic, musical, and architectural achievements, its iron, paper, and electrical equipment industries, and its scenic beautification and environmental conservation efforts".[121]

The area, which includes within its boundaries 139 properties on the National Register, including five National Historic Landmarks,[122] and four National Natural Landmarks,[123] has been considered as one of America's leading cultural resorts, and has been used for inspiration by writers, artists, musicians, and holidaymakers for more than 150 years.[124] The area, noted for its pioneering achievements in the iron,

120 For a background study of Heritage Protection in the Appalachian area, and the economic impact of Heritage Areas in the region, including the Rivers of Steel Area, see SCHOEN, R. "*Confronting the Appalachian breakdown: historic preservation law in Appalachia and the potential benefits of historic preservation for rural communities*" in 110 W. Va. L. Rev. 1303 Spring 2008.

121 Public Law 109–338 Section 272 (a) 1.

122 i) Edith Wharton's home, The Mount, Lenox, Massachusetts; ii) Herman Melville's home, Arrowhead, Pittsfield, Massachusetts; iii) W.E.B. DuBois' Boyhood Homesite, Great Barrington, Massachusetts; iv) Mission House, Stockbridge, Massachusetts; v) Crane and Company Old Stone Mill Rag Room, Dalton, Massachusetts.

123 i) Bartholomew's Cobble, Sheffield, Massachusetts, and Salisbury, Connecticut; ii) Beckley Bog, Norfolk, Connecticut; iii) Bingham Bog, Salisbury, Connecticut; and iv) Cathedral Pines, Cornwall, Connecticut.

124 Public Law 109–338 Section 272 (a) 3.

paper, and electrical generation industries, became a national leader in scenic beautification and environmental conservation efforts following the era of industrialization and deforestation maintaining a fabric of significant conservation areas.[125]

With a Native Indian presence going back over at least 10,000 years, the area was also the site of important historical events related to the American Revolution and the early civil rights movement.

The feasibility study for the area identified five thematic areas for the Heritage Area:

- **Cultural Resort** According to the feasibility study "the upper Housatonic Valley is noted for a long-standing literary tradition, the work of prominent artists and architects, and world-class music, dance, and theatre. Herman Melville wrote Moby Dick while living at Arrowhead in Pittsfield, and Nathaniel Hawthorne wrote House of the Seven Gables at Stockbridge".
- **Shaping a Scenic Landscape** One of the main features of the area is its scenic landscape. Deforested by up to 75% in early industrialisation, the area is now 75% forest. Private nature reserves were established in the 19th century, which were mainly turned into public parkland in the 20th century as new tax laws made large estates difficult to manage.
- **Cradle of Industry** The first major industry in the area was iron. In the 18th century over 40 iron furnaces were operating. Portions of six now remain. The paper industry was also important in the 19th century, and the major company Crane, still manufactures paper for US currency.
- **Revolutionary War Era and Development of Democracy: Several important events took place in the revolutionary period** The Sheffield Declaration, an early petition of grievances against British rule, was drafted at Colonel John Ashley's House in 1773, and Militias from Ethan Allen's original home in Northwest Connecticut joined the "Green Mountain Boys" in their capture of Fort Ticonderoga in 1775.
- **Other Heritage Themes: Mohican Indians, Shakers, Historic Towns** Mohicans populated the area sparsely during the era of early European settlement, and the area saw some early attempts at creating Christian Indian communities. The area also was

125 Public Law 109–338. Section 272 (a) 6.

hospitable to the Shakers, a 19th century religious sect, and the area contains an important Shaker village, Hancock. The area also has an extensive collection of historic buildings dating as far back as 1734, with many meeting houses, commons, houses, and public buildings in town centres in the area, many of which are on the National Register of Historic Places.

2 State Law

2.1 California State Law and the Protection of Historic Properties

All states have enacted laws creating state agencies with preservation responsibilities and a state register of historic places.[126] Some states have also gone further in their regulations, with substantive approaches to environmental legislation, and enabling legislation giving local authorities the State's policing powers to regulate and protect historic districts.

In this section we deal with California's regulation for the protection of Historic buildings. Relevant sections in the Public Resources Code are Historical Resources,[127] State Landmarks,[128] the Heritage Fund,[129] Archaeological sites,[130] Native American heritage[131] and the relevant sections of the California Environmental Quality Act.[132]

Additionally, the Government Code includes relevant sections on Public Records,[133] Capitol View protection,[134] the State Capital Commission,[135] the California Main Street programme,[136] elements of the Local Government Code,[137] elements of regulation of City property[138] and the Mills Act.[139]

---

126 See ZAX, L, Protection of the Built Environment: "*A Washington, D.C. Case Study in Historic Preservation*", 19 B.C. Environmental Affairs Law Review **651**, (1992).
127 California Public Resource Code 5020 –5029.
128 California Public Resource Code 5031–5033.
129 California Public Resource Code 5079.
130 California Public Resource Code 5097.1 – 5097.6.
131 California Public Resource Code 5097.9 – 5097–991.
132 California Public Resource Code 21083.2, 2084 and 21084.1.
133 California Government Code 6254.
134 California Government Code 8162.
135 California Government Code 9149.
136 California Government Code 15399.
137 California Government Code 25373 and 27288.2.
138 California Government Code 37361.
139 California Government Code 50280–50290.

Within the Health and Human Safety Code, there are sections dedicated to the removal of human remains where historic significance is suspected,[140] the treatment of hazardous buildings[141] which have historic significance, the relevant sections of the Historic Buildings Code[142] and the Marks Act on Historical Rehabilitation.

The California Civil Code regulates conservation easements,[143] while the Revenue and Taxation Code also contains historic property restrictions.[144]

Finally the Penal Code regulates the destruction, defacement of objects of archaeological or historical interest, laying out that "every person, not the owner thereof, who wilfully injures, disfigures, defaces, or destroys any object or thing of archaeological or historical interest or value, whether situated on private lands or within any public park or place, is guilty of a misdemeanour".

2.1.1 California Law implementing the National Historic Preservation Act

Californian state law implements the National Historic Preservation Act programme. It also compliments the federal historical preservation programme principally through the provisions of the Public Resource Code relevant to Historical Resources, regulating the State Historical Resources and Commission (a continuation of the State Historical Resources Commission), the California Register of Historical Resources, the State Office of Historic Preservation, the State Historical Resources Inventory, State Landmarks, the Heritage Funds, Archaeological sites, and Native American heritage. There are also important provisions in the California Environmental Quality Act.

In the same line as the Federal programme, the state programme is essentially voluntary in nature, and depends completely on receiving the consent of the owner of the property to its protection and inclusion in state programmes. In a statement of legislative intent, the legislature "recognizes that the long-term preservation and enhancement of historical resources is dependent, to a large extent, on the good will and cooperation of the general public and of the public and private owners of those resources". Given this need for good will and cooperation, state agencies are directed to interpret the law in a way

140 California Health and Human Safety Code 7050.5.
141 California Health and Human Safety Code 17922.2.
142 California Health and Human Safety Code 18950–18961.
143 Californian Civil Code 815–816.
144 California Revenue and Taxation Code 439.

designed "to elicit the cooperation of the owners of both identified and unidentified resources, to encourage the owners to perceive these resources as assets rather than liabilities".[145]

A) State Historical Resources Commission

The State Historical Resources Commission is a body composed of nine members appointed by the governor. Seven members should be recognised professionals in history, prehistoric archaeology, historic archaeology, architectural history, ethnic history, folk life, or architecture. The other two members should represent the public or "possess expertise in fields of expertise the governor deems necessary or desirable to enable the commission to carry out its responsibilities".[146]

The commission, which should meet at least four times a year, has the responsibility, amongst other things to:

- receive and evaluate applications for, and make recommendations with respect to entries on, the National Register of Historic Places to the State Historic Preservation Officer;
- conduct a state wide inventory and maintain comprehensive records of historical resources including, but not limited to, historical landmarks and points of historical interest;
- establish criteria for the recording and preservation of historical resources, and for deletions from historical registers;
- develop and adopt criteria for the rehabilitation of historic structures;
- establish policies and guidelines in compliance with state and federal requirements for a comprehensive state-wide historical resources plan which includes, but is not limited to, architecture, history, archaeology, and folk life;
- develop and update annually, based upon public hearings and active public participation, the state-wide historical resources plan;
- oversee the administration of the California Register;

145 California Public Resources code 5020.7.
146 California Public Resources code 5020.2.

- receive and evaluate applications for registration of structures, sites, or places as historical landmarks or points of historical interest;
- consult with, and consider the recommendation of, public agencies, civic groups, and citizens interested in historic preservation and develop criteria and procedures based upon public hearings and active public participation for the selection of projects to be funded through the National Historic Preservation Fund, the California Heritage Fund, and other federal and state programmes that have as their primary purpose the preservation and enhancement of historical resources.[147]

B) The California Register of Historic Resources

The California Register has the function of being an authoritative guide to California's Historic Resources,[148] to be used by public agencies, non-government organizations and private citizens, indicating which properties should be protected to a "prudent and feasible" extent from substantial adverse change.[149]

The types of property that can be included on the Register are buildings,[150] sites,[151] structures,[152] objects[153] and historic districts.[154] In

147 California Public Resources code 5020.1.

148 Includes, but is not limited to, any object, building, structure, site, area, place, record, or manuscript which is historically or archaeologically significant, or is significant in the architectural, engineering, scientific, economic, agricultural, educational, social, political, military, or cultural annals of California. California Public Resources code 5020.1 (j).

149 California Public Resources code 5024.1 (a).

150 A resource, such as a house, barn, church, factory, hotel, or similar structure created principally to shelter or assist in carrying out any form of human activity. "Building" may also be used to refer to a historically and functionally related unit, such as a courthouse and jail or a house and barn.

151 A site is the location of a significant event, a prehistoric or historic occupation or activity, or a building or structure, whether standing, ruined, or vanished, where the location itself possesses historical, cultural, or archaeological value regardless of the value of any existing building, structure, or object. A site need not be marked by physical remains if it is the location of a prehistoric or historic event, and if no buildings, structures, or objects marked it at that time. Examples of such sites are trails, designed landscapes, battlefields, habitation sites, Native American ceremonial areas, petroglyphs, and pictographs.

152 A construction made for a functional purpose rather than creating human shelter. Examples of structures include mines, bridges, and tunnels.

153 Those constructions that are primarily artistic in nature or are relatively small in scale and simply constructed, as opposed to a building or a structure. Although it may be movable by nature or design, an object is associated with a specific setting or environment. Objects should be in a setting appropriate to their significant historic use, role, or character. Objects that are relocated to a museum are not eligible for listing in the California Register. Examples of objects include fountains, monuments, maritime resources, sculptures, and boundary markers.

the case of the listing of a historic district, those individual resources contributing to its significance must also be listed. For this reason, *all individual resources* located within its boundaries must be designated as either contributing or as non-contributing to the significance of the historic district.

The criteria for inclusion on the California Register are consistent with the National Register of Historic Places criteria. However, as the regulations for the California Historic Register point out, they have been modified for state use in order to include a range of historical resources which better reflect the history of California. To be included in the Register, a Historic Resource must either:

- be associated with events that have made a significant contribution to the broad patterns of local or regional history, or the cultural heritage of California or the United States;
- be associated with the lives of persons important to local, California, or national history;
- embody the distinctive characteristics of a type, period, region, or method of construction, or represents the work of a master or possesses high artistic values; or
- have yielded, or have the potential to yield, information important to the prehistory or history of the local area, California, or the nation.[155]

The resource must also meet the criteria of "integrity", based on the UNESCO definition discussed above. Integrity or the "authenticity of an historical resource's physical identity evidenced by the survival of characteristics that existed during the resource's period of significance" is determined by such factors as retention of location, design, setting, materials, workmanship, feeling, and association. If the resource is felt to have lost integrity, then it may still be included if it maintains the potential to yield significant scientific or historical information or specific data.[156]

In general, only resources that are more that 50 years old can be considered for listing. More recent resources can only be considered if

154 Unified geographic entities which contain a concentration of historic buildings, structures, objects, or sites united historically, culturally, or architecturally. Historic districts are defined by precise geographic boundaries. Therefore, districts with unusual boundaries require a description of what lies immediately outside the area, in order to define the edge of the district and to explain the exclusion of adjoining areas.

155 California Code of Regulations, Title 14, Chapter 11.5 4852 (b) 1–2.

156 California Code of Regulations, Title 14, Chapter 11.5 4852 (c).

it can be demonstrated that sufficient time has passed to understand its historical importance. In the same line, a reconstructed building less than fifty years old may be eligible if it embodies traditional building methods and techniques that play an important role in a community's historically rooted beliefs, customs, and practices.

The Register must include all properties either included on the National Register (or considered "eligible for" the National Register), State Historical Landmarks[157] and Points of historical interest which have been reviewed by the office and recommended for listing.[158]

The Register may also include properties that have been nominated according to the administrative procedure established in the Public Resources code,[159] if they are determined significant by the commission and may include:

- individual historical resources;
- historical resources contributing to the significance of an historic district under criteria adopted by the commission;
- historical resources and historic districts designated or listed as city or county landmarks, historic properties or districts pursuant to any city or county ordinance;
- local landmarks or historic properties designated under any municipal or county ordinance;[160] and
- historical resources identified as significant in historical resources surveys.[161]

Resources identified in a Historical Resources Survey can be included in the register only if it has been carried out according to department standards, and if the property is either listed in the National Register of Historic Places, formally determined eligible for listing in the National Register, appears to be eligible for listing in the National

157 Only State Historical Landmark No. 770 and all consecutively numbered state historical landmarks are to be included obligatorily. The state office has to review eligibility for those previous to 770.

158 California Public Resources code 5024.1 (d) 3.

159 California Public Resources code 5024.1 (f).

160 California Public Resources code 5024.1 (e) 1.

161 Historical Resources Surveys are defined as "The process of systematically identifying, researching, photographing, and documenting historical resources within a defined geographic area". A Glossary of Terms as used in the California Register of Historical Resources (Title 14, Chapter 11.5, Appendix A).

Register, could become eligible for listing in the National Register or is considered to be locally significant.[162]

As in the National Register, the owner of a private property or the majority of owners for an historic district or single property with multiple owners may object to the nomination. In this case, the property cannot be listed as a historical resource in the California Register until the objection is withdrawn.[163] In the case that private property cannot be presently listed in the California Register solely because of owner objection, the property must be included in the Register as "eligible for listing".

C) The State Historic Preservation Office

The State Historic Preservation Officer is appointed by the Governor and serves as the Chief Administrative Officer of the Office of Historic Preservation and Executive Secretary of the State Historical Resources Commission. It is his or her responsibility to administer state and federally mandated historic preservation programmes,[164] fulfilling all the functions previously discussed under federal section 106 compliance.

The State Historic Preservation Office, under the guidance of the State Historic Preservation Officer, has a wide range of responsibilities, including the recommendation of properties of historical significance for nomination by the commission for the National Register of Historic Places, for registration as historical landmarks and points of historical interest, and for listing in the California Register. The office also administers state and federal incentive programmes for the preservation of historical resources. Other duties include:

- providing information on federal and state benefits for preservation projects and enhancement of historical resources;
- administering grant and loan programmes to survey historical resources, providing information and technical assistance to local, state, and national organizations to promote preservation and enhancement of historical resources by developing model

162 California Code of Regulations, Title 14, Chapter 11.5 4852 (e) 2.

163 Objections have be submitted to the commission by the owner of the private property in the form of a notarized statement certifying that the party is the sole or partial owner of the property, and that the party objects to the listing. California Public Resources code 5024.1 (f) 4.

164 National Historic Preservation Act of 1966, Section 101 (b)(3), (16 U.S.C. 470 (1966) (amended)) and Cal. Pub. Res. Code Section 5024.

ordinances, financial mechanisms, educational programmes, conferences, workshops, and other materials;

- reviewing and commenting on the impact on historical resources of publicly funded projects and programmes undertaken by other governmental agencies; and
- administering the California Heritage Fund.

With regard to policy on State-owned historic property, the State Historic Preservation Officer plays a coordinating role with the state agencies identifying and cataloguing state owned property, and, according to executive order W-26-92 "shall provide leadership to and shall coordinate the efforts of State agencies in implementing their stewardship obligations with regard to historic resources".[165]

D) The California Heritage Fund

The California Heritage Fund, born with the same principles as the National Heritage Fund, is administered by the State Historical Preservation Office. Moneys can be deposited by legislative appropriation, grants, gifts and donations.

The Fund may be used for a wide variety of purposes, including the purchase of historical resources, artefacts, or objects, or any interest herein, that are required to provide public access to cultural or historical resources. They may be acquired by purchase, option to purchase, gift, exchange, dedication, or by a combination of the above.[166]

Money in the fund may also be used for loans and grants to public agencies and non-profit organizations for historical preservation in order to further the aims of Historical Resource legislation.[167] Grants may also be made for educational purposes, and made to public agencies and non-profit organizations with the purpose of improving the management of California's historical resources which, because of natural events or human activities, have suffered impairment or loss of historic integrity. Funds may also be made available for the preparation of plans for the protection and enhancement of the historical resource.[168] Grants may also be awarded to public agencies and non-profit organizations for historical resource preservation projects that

165 Executive Order W-26-92 8th April 1992.

166 California Public Resources code 5079.21.

167 California Public Resources code 5079.25.

168 California Public Resources code 5079.51.

include the acquisition of significant archaeological resource areas and artefacts associated with those sites.[169]

2.1.2 Other Californian Laws Protecting Historic Properties

A) The Mills/Historic Properties Act of 1972

California Law, through the Historical Properties Act, allows legislative bodies to enter into contracts with private owners to restrict the use of the property in a manner which the legislative body deems reasonable. The contract should be of a minimum of 10 years and should provide for the preservation of the qualified historical property and, when necessary, to restore and rehabilitate the property to conform to the rules and regulations of the Office of Historic Preservation and the State Historical Building Code.

In order to qualify, the property must be listed in the National Register of Historic Places, be located in a registered historic district, or be listed in any state, city or county official register of historical or architecturally significant sites, places, or landmarks.[170]

The contract should provide for inspections to ensure compliance, and can be terminated by the legislature if it determines that the owner has breached any of the conditions of the contract or has allowed the property to deteriorate to the point that it no longer meets the standards for a qualified historical property, or if it determines that the owner has failed to restore or rehabilitate the property in the manner specified in the contract.[171] Cancellation of the contract can only be made after a public hearing, and in the event of cancellation the owner of the property has to pay a cancellation fee equal to 12.5% of the current fair market value of the property, as determined by the county assessor.

Entering into a contract of restriction of use has significant favourable property tax implications for an owner, given that a property is not valued on sales data on similar properties, but rather on a system based on either: a fair rent for similar properties, or the amount "a restricted historical property being valued reasonably can be expected to yield under prudent management and subject to applicable provisions under which the property is enforceably restricted".[172]

The voluntary nature of these previsions must be emphasised. The case *Prentiss v. City of S. Pasadena*, for example, (discussed below in

169 California Public Resources code 5079.61.
170 California Public Resources code 5028.1.
171 California Public Resources code 5028.4.
172 California Revenue and Taxation Code 439.2 a) 2.

the section on the California Environmental Quality Act) underlines the voluntary nature of the provisions of the Historical Properties Act in contrast to the mandatory nature of local ordinances, mandatory zoning and land use restrictions, and points out that there is nothing that obliges an owner to enter into a contract with public authorities in the California code or to apply the State Historic Building code in lieu of the normal building code.

B) Marks Historical Rehabilitation Act of 1976

The Californian legislature has also acted in the area of financing of historical rehabilitation. The Marks Historical Rehabilitation Act of 1976 recognises that "properties and structures of historical or architectural significance are an essential public resource and that it is necessary and essential that cities, counties, and cities and counties be authorized to make long-term, low-interest loans to finance the rehabilitation of properties of historic or architectural significance". The act provides local agencies with the authority to provide loans for the rehabilitation of historic properties, avoiding their deterioration "at an accelerated rate because loans from private sources are not sufficiently available for their rehabilitation".[173]

The backbone of the rehabilitation programme is the historical rehabilitation financing programme, which must be approved prior to the issuing of bonds or bond anticipation notes for historical rehabilitation financing. The plan must specify the criteria for selection of historical properties eligible for financing, including criteria of architectural[174] and historical significance.[175] The plan should also

---

173 Marks Act 37601.

174 Criteria may include, but are not limited to:
- Structures or areas that embody distinguishing characteristics of an architectural style, period, method of construction, or architectural development in a city or county;
- Notable works of a master builder, designer, or architect whose style influenced the city's or county's architectural development, or structures showing the evolution of an architect's style;
- Rare structures displaying a building type, design, or indigenous building form;
- Structures which embody special architectural and design features;
- Outstanding examples of structures displaying original architectural integrity, structurally or stylistically; or both
- Unique structures or places that act as focal or pivotal points important as a key to the character or visual quality of an area. Marks Act 37626(a).

175 Criteria may include, but are not limited to:
- Sites and structures connected with events significant in the economic, cultural, political, social, civic, ethnic, or military history of a community, state, or nation;
- Structures or areas identified with the lives of historical personages of a community, state, or nation;

specify the criteria for the selection of the historical rehabilitation area,[176] and include a map and other information on the area included.

C) California Environmental Quality Act of 1970

California's Environmental Quality Act (CEQA) is an important tool in the protection of historic properties. A major difference between this Act and NEPA is that CEQA gives substantive obligations to agency officials. The Act is applicable to all projects[177] undertaken by any state or local agency, special districts or public schools or universities. It also applies to discretionary private projects, which are defined as those that require "the exercise of judgement or deliberation by a public agency in determining whether the project will be approved, or if a permit will be issued".[178] It does not however apply where the issuance of the permit is a ministerial act.[179] In other words, CEQA

- Sites and groups of structures representing historical development patterns, including, but not limited to, urbanization patterns, railroads, agricultural settlements, and canals. Marks Act 37626 (b).

176 Criteria may include, but are not limited to:
- Areas constituting a distinct section of the city and having special character, historical, architectural, or aesthetic interest and value.
- Areas providing significant examples of architectural styles of the past, or landmarks in the history of architecture.
- Areas serving as a reminder of past eras, events, and persons important in local, state, or national history, and illustrating past living styles for future generations to observe, study, and inhabit.
- Historical and culturally significant grounds, gardens, and objects. Marks Act 37626 (c).

177 The definition of "project" must be interpreted expansively. The court in Friends of the *Sierra Railroad v. Tuolumne Park & Recreation department* reminds us that the California Supreme Court "has stated that when a court determines whether an activity is a project, the statute is to be interpreted in such manner as to afford the fullest possible protection to the environment within the reasonable scope of the statutory language." (*Friends of Mammoth v. Board of Supervisors* (1972) 8 Cal.3d 247, 259 [104 Cal. Rptr. 761, 502 P.2d 1049], disapproved on other grounds by *Kowis v. Howard* (1992) 3 Cal.4th 888, 896–897 [12 Cal. Rptr. 2d 728, 838 P.2d 250].) Activities that have been held to constitute projects include not only approvals of real estate developments, but many other decisions with a potential to affect the environment significantly.

178 California Environmental Quality Act (CEQA) and Historical Resources. California Office of Historic Preservation. Department of Parks and Recreation. Technical Assistance Series. P. 7. Examples of discretionary decisions that are given are "placing conditions on the issuance of a permit, delaying demolition to explore alternatives, or reviewing the design of a proposed project".

179 Cal. Code Regs., tit. 14, § 15369. state that "'Ministerial' describes a governmental decision involving little or no personal judgment by the public official as to the wisdom or manner of carrying out the project. The public official merely applies the law to the facts as presented but uses no special discretion or judgment in reaching a decision. A ministerial decision involves only the use of fixed standards or objective measurements, and the public official cannot use personal, subjective judgment in deciding whether or how the project should be carried out…A building permit is ministerial if the ordinance requiring the permit limits the public official to determining whether the zoning allows the structure to be built in the

does not apply to those projects that do not require discretionary judgements by public officials – that is, where the administration only plays a role of determining that certain statutory conditions have been met.[180]

The difference between "discretionary acts" and "ministerial acts" has been dealt with by the California appeals court in *Prentiss v. City of S. Pasadena,*[181] in a case where a building permit was denied for a building eligible for listing on the National Register of Historic Places. In this case the court decided that CEQA did not apply, because issuance of a building permit free of the historical architectural conditions was a ministerial act. This was due to the fact that the application for a building permit required no variance or conditional use permit and fully complied with the Uniform Building Code, and so issuance of the requested building permit was a ministerial act to which CEQA did not apply. According to the court, "appellants fail to show that any statute or ordinance gave appellants discretion to deny the permit on historical architectural grounds. In this respect appellants misinterpreted their authority under the State Historical Building Code, failing to distinguish between laws designed to encourage voluntary historical preservation by private property owners and zoning laws designed to require historical preservation".

This decision was taken despite the fact that the building work was significant, the modification of a two story building to a three story building. However, the modifications were consistent with zoning laws, and it was undisputed that, had the building not been considered a Historic Resource, the issuance of the building permit would have been unquestionably a ministerial act. The public authority, however, erroneously argued that the State Historic Building Code applied, and that had the effect of making the decision on the project a discretionary one which brought CEQA into play, and therefore they could use CEQA to impose historical architecture conditions on the permit.

The court pointed out that the mere listing of a private property on the State Historical Resources Inventory, based on its being "eligible" for listing on the National Register of Historic Places, is not by itself a substantive restriction on the owner's use of the property. With respect to privately owned historical property, California statutes "provide

---

requested location, the structure would meet the strength requirements of the Uniform Building Code, and the applicant has paid his fee."

180 ibid. Examples are roof replacements, interior alterations to residences, and landscaping changes.

181 15 Cal. App. 4th 85.

only for the registration of state historical sites, impose no restraints on use of those sites and do not require preservation in any particular form. The only purpose of the state programme is to list and mark significant sites".[182]

Crucially, in this case, the City of South Pasadena had not enacted any ordinance which regulated the properties in the neighbourhood and had not created any historic zone,[183] and based its entire case on an interpretation of the State Historic Building Code, as an "other law" which makes respondents' project discretionary.

The Court interpreted the State Historic Building Code as a tool that does not force, but rather allows

> in the case of qualified historical buildings, alterations which otherwise would not meet the standards of the regular prevailing building code, such as the Uniform Building Code. The State Historical Building Code fits the pattern of laws designed to encourage private owners to preserve historically significant properties. The State Historical Building Code is not, however, a device by which a city or county may compel an owner to preserve the historical character of the property...The inclusion of respondents' property on the state inventory made the property eligible for the benefits of the State Historical Building Code. The requirement that every city and county building department 'shall apply' the State Historical Building Code means that they must allow the owner to use the approved alternative methods. Appellants, however, would interpret the State Historical Building Code to authorize a building and planning department to prevent any alteration to an historical structure, if the alteration is not deemed 'historically correct'.

The court therefore found no basis in law for the public authority's interpretation of the State Historic Building Code.

CEQA also applies where decisions are made by public bodies that could have an indirect impact on the environment or on historical resources, such as changes in specific plans and local codes. The sale of land including historic resources by a public agency does not trigger the CEQA process, unless development plans are presented as part of the sale agreement, even though it is highly likely that the sale will be followed by development.[184]

---

182 See this sense *Foundation for San Francisco's Architectural Heritage v. City and County of San Francisco*, supra, 106 Cal.App.3d at p. 914

183 According to the court, the city did not create an historic zone under former Government Code sections 50280 and 50281 to contract with owners for preservation. The city did not enact a regulatory historic district architectural control ordinance for this neighbourhood. (Gov. Code, § 37361). The city did not even include this property within its inventory of cultural heritage landmarks so as to bring it within the permit requirements of the city's cultural heritage ordinance.

184 See in this sense *Friends Of The Sierra Railroad v. Tuolumne Park And Recreation District* "Like any other agency action, property disposal – even of a historical resource – is not a

CEQA considers that "a project that may cause a substantial adverse change in the significance of an historical resource is a project that may have a significant effect on the environment". Therefore, any of the projects that fall under CEQA, and that are not statutorily exempted, and affect historical resources, must comply with CEQA, including private projects that require decisions based on agency discretion. The definition of historical resource in CEQA is wide. The definition includes, as one would expect, properties listed on the California Register, or determined as being eligible to be listed. Additionally,

> historical resources included in a local register of historical resources...or deemed significant pursuant to National Register criteria are presumed to be historically or culturally significant..., unless the preponderance of the evidence demonstrates that the resource is not historically or culturally significant. The fact that a resource is not listed in, or determined to be eligible for listing in, the California Register of Historical Resources, not included in a local register of historical resources, or not deemed significant pursuant to National Register criteria shall not preclude a lead agency from determining whether the resource may be an historical resource.[185]

Substantial adverse change is defined in the Historical Resource Code as "demolition, destruction, relocation, or alteration such that the significance of an historical resource would be impaired". Demolition, destruction and relocation are easily defined. "Alteration such that the significance of an historical resource would be impaired" is more difficult to define. CEQA guidelines lay out that project substantially damages a historical resource when it:

a) materially alters in an adverse manner those physical characteristics of an historical resource that convey its historical significance and that justify its inclusion in or eligibility for, inclusion in the California Register of Historical Resources;
b) materially alters in an adverse manner those physical characteristics that account for its inclusion in a local register of historical resources or its identification in an historical resources survey unless the public agency reviewing the effects of the project establishes by a preponderance of evidence that the resource is not historically or culturally significant;

---

project if too little is known about environmental impacts to which it might lead to allow those impacts to be meaningfully analyzed". 147 Cal. App. 4th 643; 54 Cal. Rptr. 3d 500; 2007 Cal. App. LEXIS 171; 2007 Daily Journal DAR 1878.

185 California Public Code CEQA 21084.1.

c) materially alters in an adverse manner those physical characteristics of a historical resource that convey its historical significance and that justify its eligibility for inclusion in the California Register of Historical Resources as determined by a lead agency for purposes of CEQA.

Substantial change can be avoided where the project will be conducted in a manner consistent with the Secretary of the Interior's Standards for the Treatment of Historic Properties.[186] In this case, the project's impact on the historical resource is not considered significant. The Secretary of the Interior's Standards address issues involved in preservation, restoration, rehabilitation and reconstruction.

Relocation of a resource may be considered as not significant if it is the only alternative to demolition, and if the new location is compatible with the original character and use of the historical resource.[187]

The California Court of appeals in the *League for Protection of Oakland's Etc. Historic Resources v. City*[188] points out that CEQA defines the "environment" to include "historic" conditions within an area which will be affected by a proposed project, and that a "project that may cause substantial adverse change in the significance of an historical resource is a project that may have a significant effect on the environment." Also, "a project will normally have a significant effect on the environment if it will disrupt or adversely affect a property of historic or cultural significance to a community or ethnic or social group." Thus, the significant impacts of a discretionary project upon a historic building must be considered in an Environmental Impact Report.[189]

In the above mentioned case, it was contended by the Oakland Heritage Alliance that a building for which the City had issued a demolition permit (the Montgomery Ward Building) qualified as a historical resource for purposes of CEQA. The city contended that as the building has not been "officially designated" as "historic property" in the National Register, by the State of California, or in any formal City register, it was not a historical resource as defined by CEQA for which an Environmental Impact Report is required.

The court pointed out that CEQA created three categories of historical resources.

186 See Appendix II.

187 14 California Code of Regulations§ 4852 (d) 1.

188 52 Cal. App. 4th 896; 60 Cal. Rptr. 2d 821; 1997 Cal. App. LEXIS 100; 97 Cal. Daily Op. Service 979; 97 Daily Journal DAR 1447.

189 *Prentiss v. City of South Pasadena* (1993) 15 Cal. App. 4th 85, 92 [18 Cal. Rptr. 2d 641]).

> First, the mandatory provision of the statute specifies that buildings 'listed in, or determined to be eligible for listing in, the California Register of Historical Resources' must in all cases be granted status as historical resources. Second, buildings 'included in a local register of historical resources, as defined in subdivision (k) of Section 5020.1, or deemed significant pursuant to criteria set forth in subdivision (g) of Section 5024.1' are presumptively historical resources unless the preponderance of the evidence demonstrates otherwise. Third, buildings which do not fall within the mandatory or presumptive categories may still be deemed historical resources at the discretion of the lead agency.

The court therefore rejected the city's contention that "nothing less than official designation of a building as historic in a recognized register suffices to trigger CEQA requirements. According to the court, the language of sections 21084.1 and 5020.1 does not demand formal listing of a resource in a national, state or local register as a prerequisite to 'historical' status". In this case the statutory language used is more expansive and flexible.[190]

The court applies the "fair argument" test, derived from section 21151 of the California Code, which requires an Environmental Impact Report on any project which "may have a significant effect on the environment." An Environmental Impact Report must be prepared "whenever it can be fairly argued on the basis of substantial evidence that the project may have significant environmental impact". In this case, the Montgomery Ward Building is deemed to be an historical resource on the basis that

---

190 The court points out that Section 21084.1 includes within the mandatory definition of historical resources all buildings either "listed in, or determined to be eligible for listing" in the California Register. Section 5020.1, subdivision (k), defines presumptively historic buildings in similarly disjunctive language by stating that a "local register of historical resources' means a list of properties officially designated or recognized as historically significant by a local government pursuant to local ordinance or resolution". The California Register description of historical resources also encompasses eligibility in addition to listing; it provides that "historical resources" shall include "California properties formally determined eligible for, or listed in, the National Register of Historic Places." (§ 5024.1, subd. (d)(1)). Section 21084.1 provides further indication that official designation is not the sole qualifying standard by stating that even those resources not listed or determined to be eligible in the state or local register or survey may nevertheless be classified as historical by a lead agency. Even the City's historic preservation element takes a consistent approach; it provides: "For purposes of environmental review under the California Environmental Quality Act, any change that has the potential to disqualify an existing or Potential Designated Historic Property from Landmark or Preservation District eligibility or may have substantial adverse effects on the property's Character-Defining Elements will normally, unless adequately mitigated, be considered to have a significant effect." And finally, if historical resources were limited to properties actually listed, owner resistance to inclusion or mere government inaction might forestall preparation of an Environmental Impact Report for a worthy structure, a result certainly not sanctioned by CEQA. (See *Orinda Assn. v. Board of Supervisors* (1986) 182 Cal. App. 3d 1145, 1153, fn. 3 [227 Cal. Rptr. 688].)

> the City's own internal documentation consistently recognized the historical significance of the Montgomery Ward Building. It was described in the official City historical survey as at least arguably, if not definitively, eligible for inclusion in the National Register of Historic Places. The Oakland City Historical Survey was an elaborate, comprehensive survey authorized and conducted by the City as part of the general plan for purposes of 'environmental review' and definitive classification of buildings as historic landmarks. It was not merely an information gathering endeavour for general classification and inventory purposes.

An Environmental Impact Report can be set aside by the court if an agency has prejudicially abused its discretion. While much deference must be given to an agency's discretion under CEQA, abuse of discretion is established if the agency has not proceeded in a manner required by law or if the determination is not supported by substantial evidence.[191] The substantial evidence standard must be applied to conclusions, determinations and evidence.[192]

An Environmental Impact Report must consider alternatives and mitigations to a project where damage to the environment, including historic resources, is a factor. The alternatives and mitigations must be "feasible", which is defined under CEQA as "capable of being accomplished in a successful manner within a reasonable period of time, taking into account economic, environmental, social, and technological factors".[193] Feasibility of alternatives must, of course, be considered in relation to the costs of the project in itself.

Another case, which involved the Jackling House, a Spanish colonial revival building designed by George Washington Smith in the 1920s,[194] illustrates what needs to be taken into account in an environmental impact review. In this case *Uphold Our Heritage v. Town of Woodside*,[195] the Town of Woodside authorized the demolition of the house, despite contrary staff recommendations, in order to make

---

191 Substantial evidence is defined as "enough relevant information and reasonable inferences from this information that a fair argument can be made to support a conclusion, even though other conclusions might also be reached." "Substantial evidence shall include facts, reasonable assumptions predicated upon facts, and expert opinion supported by facts". See (*Bakersfield Citizens for Local Control v. City of Bakersfield* (2004) 124 Cal.App.4th 1184, 1198 [22 Cal. Rptr. 3d 203].)

192 Section 1094.5 of the Code of Civil Procedure establishes that "... the court shall not exercise its independent judgment on the evidence but shall only determine whether the act or decision is supported by substantial evidence in the light of the whole record".

193 § 21061.1; see CEQA Guidelines, § 15364.

194 The mansion was built in 1925 for Daniel Jackling, who was a key figure in the American copper industry and contains many unique copper fixtures reflective of Jackling's work in the mining industry.

195 147 Cal. App. 4th 587; 54 Cal. Rptr. 3d 366; 2007 Cal. App. LEXIS 165; 2007 Daily Journal DAR 1818.

way for a smaller family residence, finding that the proposed alternatives to the demolition identified in an environmental impact report were not feasible and that overriding considerations justify approval of a conditional demolition permit despite the adverse impact on the environment. The court struck down the town's decision to allow demolition because the Environmental Impact Report did not sufficiently compare the cost of the alternatives to the proposed project, in such a way as to allow the town councils "overriding considerations" defense.

The Jackling House – front porch. Courtesy of Uphold Our Heritage, Woodside, California.

The Jackling House – front view Courtesy of Uphold Our Heritage, Woodside, California.

The Environmental Impact Report had proposed five alternatives:

1. **No Project Alternative** This alternative would have involved the withdrawal of the project being analysed, with the resulting impacts generally being a continuation of the existing conditions on the project site.
2. **Historic Rehabilitation of the Jackling House** This alternative would have involved rehabilitation and restoration of the Jackling house to allow for the potential habitation of the home maintaining the structure's historic significance. All character-defining features, finishes, and spaces would have been retained while allowing for upgrades and changes to the kitchens and bathrooms, with an estimated cost of approximately $4.9 million.
3. **Historic Rehabilitation of the Jackling House and New Addition** This alternative would have involved the

rehabilitation of the Jackling house with modifications to the existing design to create a more conventional floor plan, especially to the second floor. This alternative had an estimated cost of approximately $9.0 million.

4. **Onsite Relocation and Historic Rehabilitation of the Jackling House** This alternative would have included relocating the Jackling house to another portion of the project site and rehabilitating the house with an estimated cost to implement of approximately $6.6 million.
5. **Offsite Relocation and Historic Rehabilitation of the Jackling House** This alternative included relocating the Jackling house to an unknown off-site location and rehabilitating the house and would cost about $6.0 million to rehabilitate the house and additional site work.

The staff preparing the Environmental Impact Report forwarded it to the planning commission recommending that permission to demolish be denied. Nevertheless, the commission decided that these alternatives were not feasible, and that, while demolition of the Jackling House would have significant unavoidable environmental impacts, the Town's interest in conserving its open space resources outweighed the impacts to the historic resource, and issued a permit for demolition. The court reversed this decision on the basis that

> there was no evidence of any economic analysis whatsoever to compare the cost of the proposed alternatives (which costs are estimated in the Environmental Impact Report, except for alternative number 5) versus the cost of the proposed project, i.e., the estimated cost of the new residence, because no costs of building the new residence were provided to the Town Council because the owner had declined to provide any design, plans, or specification of the new residence until after his demolition permit was granted.

The Court pointed out that while according to the Supreme Court's previous decision, *City of Marina v. Board of Trustees of California State University,* "when a public agency has found that a project's significant environmental effects cannot feasibly be mitigated, the agency may nevertheless proceed with the project if it also finds that specific overriding economic, legal, social, technological, or other benefits of the project outweigh the significant effects on the environment", in this case, "since the record does not support the Council's finding that all of the alternatives included in the Environmental Impact Report are infeasible, the Council's statement of overriding circumstances is necessarily invalid". The court, citing the previous *City of Marina* case, points out that CEQA

> does not authorize an agency to proceed with a project that will have significant, unmitigated effects on the environment, based simply on a weighing of those effects against the project's benefits, unless the measures necessary to mitigate those effects are truly infeasible. Such a rule, even were it not wholly inconsistent with the relevant statute, would tend to displace the fundamental obligation of each public agency to mitigate or avoid the significant effects on the environment of projects that it carries out or approves whenever it is feasible to do so.

The Court also pointed out that the feasibility of an alternative cannot only be based on the difference in cost between alternatives, but rather that

> economic unfeasibility is not measured by increased cost or lost profit, but upon whether the effect of the proposed mitigation is such that the project is rendered impractical. The fact that a project costs too much to be profitable or cannot operate at a profit so as to render it impractical does not hinge on the wealth of its proponent. No proponent, whether wealthy or not, is likely to proceed with a project that will not be economically successful. But, if the project can be economically successful with mitigation, then CEQA requires that mitigation, regardless of the proponent's financial status.

Part of the Town's defense also rested on the contention that the alternatives to the demolition project were unfeasible because, under relevant conservation laws, it lacked the power to force the present owner to restore the property, and that therefore, the alternatives suggested were "not legally feasible". This led them to the conclusion that only feasible alternatives are those that the owner is willing to accept. The Court rejected this argument, stating that "any such finding would not be defensible". Appellants contend that because the Town are not able to compel the owner to restore the Jackling House or to sell his property, the alternatives "are not feasible because they are not capable of being accomplished in a successful manner…The willingness of the applicant to accept a feasible alternative, however, is no more relevant than the financial ability of the applicant to complete the alternative. To define feasible as appellants suggest would render CEQA meaningless".[196]

It is important to point out the limitations on how CEQA does protect historic properties based on this case, as nothing in the case obliges the owner to restore or maintain the Jackling House. In fact, the only thing the court's decision achieves is the striking down of the

196 The owner and town appealed to the Supreme Court, where petition for review was denied.

Town's decision to allow demolition only while the record does not show that alternatives are not feasible. The preservation of the house continues to depend on the owner's willingness to actively preserve it. This is in direct contradiction to the European legal systems studied in this publication, in which the relevant public authorities would have many possible alternatives open to force the owner to take action to protect and maintain the building, as well as the possibility to carry out such actions themselves, passing on the costs to the owner at a later date.

3 Conservation Easements in Federal and State Law

Conservation easements are an important tool in the protection of North America's cultural Heritage. It is a voluntary tool that creates a deed that will permanently be connected with a property, and can affect as many elements of the property as the owner wishes. Easements on properties included in the National Historic Register qualify for important tax credits and ensure that the property is protected permanently, even once a property is sold on. This is especially important given the voluntary nature of the protection of historic properties in the United States, essentially at the mercy of their owners.

An easement can be defined as a right held by one person to make specific, limited use of land owned by another person, and is in many senses compatible with the Spanish legal concept of "servidumbre". A conservation easement is, in effect, a legal agreement between the legal owner of a property, and a recognised body, that is designed to protect important conservation elements of that property permanently.

Conservation easements have become an important technique in the protection of property from development and alteration in recent years as an alternative to the outright purchase of land by a non-profit organization or by government bodies given the rise in land prices, and the relative cost effectiveness of the purchase of an easement. One of the most important features of a preservation easement is the tax deductions available, regulated in the IRS regulations.[197] A qualified conservation contribution is the contribution of a qualified real property interest to a qualified organization exclusively for conservation purposes. Crucially, to be eligible for a deduction, the conservation purpose must be protected in perpetuity.

---

197 (26 C.F.R. § 1.170A–14).

In order to be eligible for deduction, the following conditions must be met:

- The donation must be perpetual. According to the IRS regulations, a "perpetual conservation restriction", eligible for deduction, is a qualified real property interest. A "perpetual conservation restriction" is a restriction granted in perpetuity on the use which may be made of real property – including, an easement or other interest in real property that under state law has attributes similar to an easement (e.g., a restrictive covenant or equitable servitude). The terms easement, conservation restriction, and perpetual conservation restriction are understood to have the same meaning;
- The donee must be a qualified organization, have a commitment to protect the conservation purposes of the donation, and have the resources to enforce the restrictions;[198]
- The instrument of conveyance must prohibit the donee from subsequently transferring the easement unless the donee organization, as a condition of the subsequent transfer, requires that the conservation purposes which the contribution was originally intended to advance continue to be carried out;[199]
- The donation must be for the preservation of a historically important land area or a certified historic structure;[200]
- The donation must allow for some visual access to the property;
- The donation must be exclusively for conservation purposes;
- The donation must be enforceable in perpetuity, subject to legally enforceable restrictions, (for example, by recordation in

---

198 CFR 170 A14 c 1.

199 CFR 170 A14 c 2.

200 Historically important land areas are defined as: an independently significant land area including any related historic resources (for example, an archaeological site or a Civil War battlefield with related monuments, bridges, cannons, or houses) that meets the National Register Criteria; any land area within a registered historic district including any buildings on the land area that can reasonably be considered as contributing to the significance of the district; and any land area (including related historic resources) adjacent to a property listed individually in the National Register of Historic Places (but not within a registered historic district) in a case where the physical or environmental features of the land area contribute to the historic or cultural integrity of the property. A certified historic structure is defined as any building which is either listed in the National Register or located in a registered historic district. IRC § 170(h)(4)(A)(iv).

the land records of the jurisdiction in which the property is located);

– The donations must meet the requirements of state law.

The valuation of the easement donation is complex, although in its simplest terms it is based on the fair market value of a property before and after the conservation easement. The IRS regulations give the following example: C owns Greenacre, a 200-acre estate containing a house built during the colonial period. At its highest and best use, for home development, the fair market value of Greenacre is $300,000. C donates an easement (to maintain the house and Green acre in their current state) to a qualifying organization for conservation purposes. The fair market value of Greenacre after the donation is reduced to $125,000. Accordingly, the value of the easement and the amount eligible for a deduction is $175,000 ($300,000 less $125,000). Given the complexities of the regulations however, the value of an easement must be determined by a "qualified appraiser".

Although the deduction system is complex, generally allowances for conservation easements made after 2006 are 50 percent of a taxpayer's contribution base for the year of the donation, with any excess allowed to be carried over for up to five additional years.

Preservation easements have been reformed in the Pension preservation Act of 2006,[201] with new conditions intending to prevent abuse of the system. Section 1213 places restrictions on easements with respect to buildings in registered historic districts, establishing that:

– They must include a restriction which preserves the entire exterior of the building (including the front, sides, rear, and height of the building);
– They must prohibit any change in the exterior of the building which is inconsistent with the historical character of the building;
– Both the owner and the donee receiving the covenant must enter into a written agreement certifying that the donee is a qualified organization with a purpose of environmental protection, land conservation, open space preservation, or historic preservation, and has the resources to manage and enforce the restriction and a commitment to do so; and

201 Pub. L. No. 109–280, 120 Stat. 780 (2006).

- The Internal Revenue Service must be given detailed information including photographs of the entire exterior of the building, and a description of all restrictions on the development of the building.

The section of the Californian civil code that regulates easements considers that "the preservation of land in its natural, scenic, agricultural, historical, forested, or open-space condition is among the most important environmental assets of California" and therefore "declares it to be the public policy and in the public interest of this state to encourage the voluntary conveyance of conservation easements to qualified non-profit organizations".[202]

The civil code defines a conservation easement as

> any limitation in a deed, will, or other instrument in the form of an easement, restriction, covenant, or condition, which is or has been executed by or on behalf of the owner of the land subject to such easement and is binding upon successive owners of such land, and the purpose of which is to retain land predominantly in its natural, scenic, historical, agricultural, forested, or open-space condition.[203]

Conservation easements may only be acquired by tax-exempt non-profit organizations whose primary purpose is the preservation, protection, or enhancement of land in its natural, scenic, historical, agricultural, forested, or open-space condition or use, or by the state or any city, county, city and county, district, or other state or local governmental entity, if otherwise authorized to acquire and hold title to real property and if the conservation easement is voluntarily conveyed.[204]

Any other uses of the property not included in the easement are completely unaffected "including the right to engage in all uses of the land not affected by the easement nor prohibited by the easement or by law".[205]

The California civil code lays out that a conservation easement is an interest in real property voluntarily created and freely transferable in whole or in part. It is also has a permanent effect on the property, and is deemed to be an interest in real property notwithstanding the fact that it may be negative in character. The nature of the conservation

---

202 California Civil Code 815.
203 California Civil Code 815.1.
204 California Civil Code 815.3.
205 California Civil Code 815.4.

easement must be laid out specifically in the instrument creating or transferring the easement.[206]

Conservation easements are widely used in the United States for historic preservation. In testimony to Congress, the National Trust for Heritage Protection points out that, although this national organisation, chartered by Congress to promote public participation in the preservation of America's heritage, does hold over 100 conservation easements, ranging from "archaeological ruins in rural Colorado, open farmland next to our historic sites in Virginia, a modest but unaltered Cape Cod saltbox cottage in Massachusetts, a modernist classic in California, a frontier fort in Texas, and a number of important National Historic Landmark structures from Florida to Oregon",[207] the majority of conservation easements are held by local and state organisations.

The National Park Service directory of organisations includes seven national organisations qualified to accept easements, and over one hundred more working on a state and local level.[208]

4 The Protection of Historic Properties in Local Historic Preservation Ordinances and Landmark Laws

4.1 Local Ordinances

Scholars are usually surprised by the lack of protection given at Federal and State level to historic properties. While the National Historic Preservation Act does set up the National Register, it deals essentially with federally owned properties and the effects on historic properties caused by undertakings by federally managed projects. Additionally, given the widespread use of the National Register by other statutes, Congress in 1980 mandated that the owner must concur with the placing of his/her property on the Register. In principle, European scholars are generally surprised that nothing in Federal Law prohibits a private owner from disposing of a historic property as they wish, including its demolition.

The key to this, according to Mayes, lies in the complex governmental layering of the United States, where the Federal government exists on a grant of authority from the people to the

206 California Civil Code 815.2.

207 Testimony Before the Committee on Ways & Means Subcommittee on Oversight Hearing on Tax Deductions for Façade Easement Donations June 23, 2005.

208 Historic Preservation Easements: A Directory Of Historic Preservation Easement Holding Organizations. National Park Service 2003.

national government. Only those powers that were specifically given to the nation are powers of the national government. All others are reserved to the states or to the people. The legal sovereignty of the states still exists today, and the local system is created by a grant of power from the states to the local governments. For this author, the National Historic Preservation Act is a "very narrow part of the strategy for dealing with Historic Preservation in the United States. It is only one of the components in the list of various protective measures that should be a part of any programme of historic preservation".[209]

As previously seen, the protection by individual states of historic resources involves, as well as the administration of the federal programme, amongst other measures, tax credits for historical preservation, regulation of easements, and environmental quality acts. Rothstein takes the point of view that an important part of state legislation is enabling legislation,

> which confers on municipalities the police power of the state to regulate the preservation of historic or architecturally significant private property. This enabling legislation grants municipalities the authority to establish local landmark commissions and the power to enact preservation ordinances protecting these properties as historic districts and individual landmarks. This type of legislation is more important than registration programs for protecting historic or architecturally significant structures because it allows municipalities to require private property owners to preserve their properties.[210]

The National Historic Preservation Act also allows local governments to become involved with the administration of the federal programme through a certification process that can be carried out either by the State Historic Preservation Officer, or by the Secretary of the Interior if the local government:

- enforces appropriate State or local legislation for the designation and protection of historic properties;
- has established an adequate and qualified historic preservation review commission by State or local legislation;
- maintains a system for the survey and inventory of historic properties;

---

209 MAYES, T. "*Impact Assessments as a Tool for Protection against the State*". In Legal and Financial Aspects of Architectural Conservation: The Smolenice Castle Conference. Dundurn Press 1997. P. 60.

210 ROTHSTEIN, S "Takings Jurisprudence Comes In From the Cold: Preserving Interiors Through Landmark Designation" in 26 Conn. L. Rev. 1105.

- provides for adequate public participation in the local historic preservation programme, including the process of recommending properties for nomination to the National Register; and
- satisfactorily performs the responsibilities delegated to it under this Act.[211]

Many states also regulate the creation of Local Historic Districts (LHDs), as is the case of Connecticut, where LHDs that include diverse forms of protection for historic resources can be established after a referendum where two thirds of property owners within the proposed district agree to its creation.

It is however, with the creation of Local Historic Preservation Ordinances, that local governments make their most important contribution to Historic Preservation. As Miller[212] points out many decisions about local historic resources often take place at local town councils. Therefore, in addition to being aware of federal and state laws that help preserve historic resources, preservationists should understand local law. Kourtis emphasises the importance of local government, emphasising their role in historic preservation.[213] While not diminishing the importance of local ordinances, it is worth pointing out that, following the general lines of US Historic Preservation, they tend only to protect exteriors of buildings. Rothstein[214] points out that according to surveys, less than ten percent of local ordinances include any provisions for protecting the interiors of buildings, and that most statutes and ordinances pertain specifically to building exteriors.

In California, cities and counties are authorized to enact ordinances and regulations for the protection of historic places or districts.[215] In *Bohannan v. City of San Diego*, for example, the court upheld a San Diego architectural control ordinance based on Government Code 37361, which regulated buildings in the historic area known as "Old Town".[216]

---

211 NHPA 101 c.

212 MILLER, J. "A Layperson's Guide to Historic Preservation Law: A Survey of Federal, State, and Local Laws Governing Historic Resource Protection", in 1 Historic Preservation Law 3 (ALI-ABA Committee on Continuing Professional Education, 1999), p. 9.

213 PAPADEMETRIOU KOURTIS, J. "The Constructive Trust: Equity's Answer to the Need for a Strong Deterrent to the Destruction of Historic Landmarks", 16 B.C. Envtl. Aff. L. Rev. 793, p. 804, 1989.

214 ROTHSTEIN, S. "Takings Jurisprudence Comes In From the Cold: Preserving Interiors Through Landmark Designation" in 26 Conn. L. Rev. 1105.

215 Government Code sections 37361 and 25373.

216 (1973) 30 Cal.App.3d 416, 420 [106 Cal.Rptr. 333].

4.2 Local Historic Preservation Ordinances and the Takings Question

Local Ordinances typically place restrictions on the owner's right to alter or demolish properties deemed to be relevant to the historic and cultural heritage of the locality, although they do tend only to protect the exterior of the properties and not the interior. They therefore restrict the rights of owners of private property to dispose of that property as they wish, without any monetary compensation to them, which seems to clash with the constitutional rights of owners which prohibit the taking of property without just compensation, therefore violating the Fifth Amendment,[217] made applicable to states through the fourteenth[218] amendment to the US constitution.

The Supreme Court addressed these concerns in the landmark 1978 case *Penn Central Transportation Co. v. New York City.*[219] This case dealt with New York City's Landmarks Preservation Law (Landmarks Law),[220] "which was enacted to protect historic landmarks and neighbourhoods from precipitate decisions to destroy or fundamentally alter their character". Under this law, a Landmarks Preservation Commission (Commission) may designate a building to be a

217 No person shall be...deprived of life, liberty, or property, without due process of law; nor shall private property be taken for public use, without just compensation.

218 No state shall make or enforce any law which shall abridge the privileges or immunities of citizens of the United States; nor shall any state deprive any person of life, liberty, or property, without due process of law; nor deny to any person within its jurisdiction the equal protection of the laws.

219 438 US 104 *Penn Central Transportation Co. v. New York City* APPEAL FROM THE COURT OF APPEALS OF NEW YORK No. 77–444 Argued: April 17, 1978 – Decided: June 26, 1978. McLamb defines succinctly the importance of the Penn Central decision "First, the Supreme Court found historic preservation to be a valid governmental goal and put to rest concerns over the legitimacy of governmental property restrictions. Second, preservation programmes were catalyzed by the Court's decision that a historic preservation law, which limited alterations to property established as landmarks and historic district, was valid to achieve historic preservation. Third, the Court focused the takings inquiry on the entire property interest, not only on the part directly affected, and held that property restrictions are legitimate only if the owner is granted a 'reasonable return' on investment. Thus, the Court ruled that a property owner must be denied all use of his or her property to establish a regulatory taking. Finally, the ruling stood for the premise that property owners may not receive or be entitled to the maximum use of their property. According to the Supreme Court, the submission that property owners may establish a 'taking' simply by showing that they have been denied the ability to exploit a property interest that they theretofore had believed was available for development is quite simply untenable". MCLAMB, S Preservation law survey 2001: State preservation law in 8 Wid. L. Symp. J. 463.

220 N. Y. C. Admin. Code, ch. 8-A, 205-1.0 et seq. (1976). For a comprehensive review of New Yorks Landmark Law see S. Rothstein "Takings Jurisprudence Comes in from the Cold: Preserving Interiors through Landmark Designation" in 26 Conn. L. Rev. 1105.

"landmark" on a particular "landmark site" or may designate an area to be a "historic district." The law empowers the Board of Estimate to "modify or disapprove the designation, and the owner may seek judicial review of the final designation decision". Designation requires that the owner must keep the building's exterior "in good repair" and before exterior alterations are made must secure Commission approval. New York zoning law also allows for "Transferable Development Rights" for properties which have had their uses restricted.

Grand Central Terminal, owned by Penn Central Transportation Co, was designated a Landmark and the block it occupied a landmark site. When plans to redevelop the area together with UGP holdings were refused by the Commission as "destructive of the Terminal's historic and aesthetic features", the decision was appealed on the grounds that "the application of the Landmarks Law had 'taken' their property without just compensation in violation of the Fifth and Fourteenth Amendments and arbitrarily deprived them of their property without due process of law in violation of the Fourteenth Amendment".

It is important to point out that the appellants did not contest the New York City's objective of preserving structures and areas with special historic, architectural, or cultural significance was an entirely permissible governmental goal. Indeed, the Supreme Court in previous judgements had declared that "States and cities may enact land-use restrictions or controls to enhance the quality of life by preserving the character and desirable aesthetic features of a city".[221]

The appellants accepted both that the parcel of land occupied by Grand Central Terminal must, in its present state, be regarded as capable of earning a reasonable return, and that the Transferable Development Rights afforded appellants by virtue of the Terminal's designation as a landmark are valuable, even if not as valuable as the rights to construct above the Terminal. What was therefore at issue was the appellant's view that the restriction of property use constitutes a "taking".[222] Appellants therefore sought injunctive relief barring the city from using the Landmarks Law to impede the construction of any structure that might otherwise lawfully be constructed on the Terminal site, and damages for the "temporary taking". The court characterized

221 *New Orleans v. Dukes*, 427 US 297 (1976); *Young v. American Mini Theatres*, Inc., 427 US 50 (1976); *Village of Belle Terre v. Boraas*, 416 US 1, 9 –10 (1974); *Berman v. Parker*, 348 US 26, 33 (1954); *Welch v. Swasey*, 214 US, at 108.

222 The "takings" question is well summarised by Chapman in "To save and save not: the historic preservation implications of the property rights movement" in 77 B.U.L. Rev. 111 Feb.

the appellants position thus: “Stated baldly, appellants’ position appears to be that the only means of ensuring that selected owners are not singled out to endure financial hardship for no reason is to hold that any restriction imposed on individual landmarks pursuant to the New York City scheme is a ‘taking’ requiring the payment of ‘just compensation.’ Agreement with this argument would, of course, invalidate not just New York City’s law, but all comparable landmark legislation in the Nation”.

The court therefore had to decide whether the restrictions placed on the development of the property constituted an unjust taking within the meaning of the Fifth Amendment. The court pointed out that, of course, taking was more easily identifiable where there has been a physical invasion of property, and more complicated where the taking comes about due to interference which “arises from some public programme adjusting the benefits and burdens of economic life to promote the common good”. The earlier decision of *Pennsylvania Coal Co. v. Mahon*[223] had found that where a state statute that substantially furthers important public policies so frustrates distinct investment-backed expectations as to amount to a “taking”, compensation for that taking is required.

The court points out that where state courts have found that the prohibition of particular uses of land is beneficial to “health, safety, morals and general welfare” the court has upheld land-use regulations that destroyed or adversely affected recognized real property interests. The court recognizes that the establishment of taking is a difficult issue to define.

According to the court:

> the question of what constitutes a ‘taking’ for purposes of the Fifth Amendment has proved to be a problem of considerable difficulty. While this Court has recognized that the ‘Fifth Amendment’s guarantee...[is] designed to bar Government from forcing some people alone to bear public burdens which, in all fairness and justice, should be borne by the public as a whole’,[224] this Court, quite simply, has been unable to develop any ‘set formula’ for determining when ‘justice and fairness’ require that economic injuries caused by public action be compensated by the government, rather than remain disproportionately concentrated on a few persons.[225] Indeed, we have frequently observed that whether a particular restriction will be rendered

223 *Pennsylvania Coal Co. v. Mahon* 260 US 393 (1933) (1922).

224 *Armstrong v. United States*, 364 US 40 (1960).

225 See *Goldblatt v. Hempstead*, 369 US 590, 594 (1962).

invalid by the government's failure to pay for any losses proximately caused by it depends largely 'upon the particular circumstances [in that] case'.[226]

The court rejected the appellant's position on three fronts. In the first place, a landmark law does not discriminate against individual owners differently than a zoning law.

In second place, the court rejects the position that a decision to designate a structure as a landmark "is inevitably arbitrary or at least subjective, because it is basically a matter of taste," thus unavoidably singling out individual landowners for disparate and unfair treatment. The court points out that this decision is subject to judicial review (not sought in this case), and that courts do not have difficulties dealing with this sort of subjective decision making.

In third place, and crucially, the court rejects the appellants contention that "New York City's law is inherently incapable of producing the fair and equitable distribution of benefits and burdens of governmental action which is characteristic of zoning laws and historic-district legislation and which they maintain is a constitutional requirement if 'just compensation' is not to be afforded". Legislation designed to promote the general welfare commonly burdens some more than others, and this does not invalidate it

> in any event, appellants' repeated suggestions that they are solely burdened and unbenefited is factually inaccurate. This contention overlooks the fact that the New York City law applies to vast numbers of structures in the city in addition to the Terminal – all the structures contained in the 31 historic districts and over 400 individual landmarks, many of which are close to the Terminal. Unless we are to reject the judgement of the New York City Council that the preservation of landmarks benefits all New York citizens and all structures, both economically and by improving the quality of life in the city as a whole – which we are unwilling to do – we cannot conclude that the owners of the Terminal have in no sense been benefited by the Landmarks Law.

The court noted that the economic losses suffered by the appellants is in some sense mitigated by the imperfect system of Transferable Development Rights, and that it is not entirely clear that any development of the property has been prohibited.

It is worth taking note of the dissenting opinion which does differentiate between zoning laws and landmark legislation. According to the dissenting judges

226 *United States v. Central Eureka Mining* Co., 357 US 155, 168 (1958); see *United States v. Caltex*, Inc., 344 US 149, 156 (1952).

> the record is clear that the proposed addition to the Grand Central Terminal would be in full compliance with zoning, height limitations, and other health and safety requirements. Instead, appellees are seeking to preserve what they believe to be an outstanding example of beaux arts architecture. Penn Central is prevented from further developing its property basically because too good a job was done in designing and building it. The city of New York, because of its unadorned admiration for the design, has decided that the owners of the building must preserve it unchanged for the benefit of sightseeing New Yorkers and tourists.

According to the dissenting opinion, unlike land-use regulations, the City of New York does not merely prohibit Penn Central from using its property in a narrow set of noxious ways, but rather place an affirmative duty on Penn Central to maintain the Terminal in its present state and in "good repair." Therefore the appellants are not free to use their property as they see fit within broad outer boundaries,

> but must strictly adhere to their past use except where appellees conclude that alternative uses would not detract from the landmark. While Penn Central may continue to use the Terminal as it is presently designed, appellees otherwise 'exercise complete dominion and control over the surface of the land'. Noting that in *United States v. Dickinson*[227] the court declared that property is taken in the constitutional sense when inroads are made upon an owner's use of it to an extent that, as between private parties, a servitude has been acquired.

The dissenting judges felt that the appellees should compensate the owner for its loss. A taking does not become a no compensable exercise of police power simply because the government in its grace allows the owner to make some "reasonable" use of his property, stating that "it is the character of the invasion, not the amount of damage resulting from it, so long as the damage is substantial, that determines the question whether it is a taking".

4.3 The Historic Preservation Ordinance of the City of Phoenix

A good example of how local ordinances are used to protect historic properties on a local level, and the administrative procedure involved, is the Historic Preservation Ordinance passed by the city of Phoenix, Arizona.

4.3.1 Aims of the Historic Preservation Ordinance

The Historic Preservation Ordinance of Phoenix has a declared purpose of promoting a public policy that protects, enhances and preserves properties and areas of historical, cultural, archaeological

227 331 US 745, 748.

and aesthetic significance, given that this preservation "is in the interests of the health, prosperity and welfare of the people of the city of Phoenix".[228] The city government takes note that it must recognize "past needless losses of historic properties which had substantial value to the historical and cultural heritage of the citizens of Phoenix".

The ordinance creates a Historic Preservation Commission, a Historic Preservation Officer, the Phoenix Historic Register and administrative procedures for the creation of Historic Preservation Districts.

The ordinance declares that the creation of the figure of the Historic Preservation District aims to "effect and accomplish the protection, enhancement and preservation of improvements and landscape features of landmarks, districts and archaeological resources which represent distinctive elements of the City's cultural, educational, social, economic, political, architectural and archaeological history", as well as safeguarding "the City's historic, aesthetic and cultural heritage, as embodied and reflected in such districts".

Amongst other aims of the ordinance is that of fostering "civic pride in the accomplishments of the past, to protect and enhance the City's attraction to visitors and the support and stimulus to the economy thereby provided and to promote the use of Historic Preservation Districts and properties for the education, pleasure and welfare of the people of the City of Phoenix".

Within those Historic Preservation Districts, the ordinance aims to "retain and enhance those properties which contribute to the character of the Historic Preservation District and to encourage their adaptation for current use" and to "assure that alterations of existing structures are compatible with the character of the Historic Preservation District". Essential to this task is assuring that new construction is compatible with the character of Historic Preservation District.

4.3.2 Administrative Procedure for Designation of Historic Properties and Districts

The application for the creation of a Historic Preservation District should follow the administrative procedures set out in Phoenix's zoning laws,[229] and is then passed on to the Historic Preservation Officer of the City of Phoenix. The officer then passes the application

228 Historic Preservation Ordinance City of Phoenix 802 A.

229 Section 506 Phoenix Zoning Ordinance.

on to the Historic Preservation Commission, together with a summary of the project and recommendations.

The Historic Preservation Commission must then evaluate the property proposed to be zoned as Historic District for a demonstrated quality of significance in local, regional, state or national history, architecture, archaeology, engineering or culture. The criteria used are significance,[230] integrity[231] and age.[232]

The Historic District should be delimited in such a way as to ensure that the district contains documented historic, architectural, archaeological or natural resources, that the district boundaries coincide with documented historic boundaries such as early roadways, canals, subdivision plats or property line, and that other, non-historic resources or vacant land is included where necessary to create appropriate boundaries. After another, more formal, public hearing, the city council makes a decision on the zoning application according to the zoning ordinance.

4.3.3 Effects of Designation

Once a property is declared as a historic property then the ordinance prohibits that any building, permanent sign, or other structure within an Historic Preservation District may be erected, demolished, moved, restored, rehabilitated, reconstructed, altered or changed in exterior appearance until plans for such activities have been submitted to and approved by the Historic Preservation Officer, Historic Preservation Commission or City Council, and a Certificate of No Effect,[233] a Certificate of Appropriateness,[234] or a Demolition Approval is issued.

---

230 Associated with the events that have made significant contribution to the broad patterns of our history; and/or b) Associated with the lives of persons significant in our past; and/or c) Embody the distinctive characteristics of a type, period or method of construction or that represent the work of a master or that possess high artistic values or that represent a significant and distinguishable entity whose components may lack individual distinction; and/or d) Have yielded or may be likely to yield information important in the understanding of our pre-history or history of the City of Phoenix.

231 Retain sufficient integrity of location, design, setting, materials, workmanship, feeling and association to convey their significance.

232 Are a) at least fifty years old; or b) Have achieved significance within the past fifty years if the property is of exceptional importance.

233 A certificate of no effect can be issued where: a) It is determined the proposed work is minor and clearly within adopted design guidelines, and b) If modifications to the proposed work are requested by the HP Officer, they are agreed to by the applicant, and c) In any case, the proposed work will not diminish, eliminate, or adversely affect the historic character of the subject property or its affect on the district.

234 The certificate of appropriateness, required if a certificate of no effect is not issued, is subject to public hearing and to the following standards. 1. The proposed work will be compatible with the relevant historic, cultural, educational or architectural qualities characteristic of the

This prohibition does not prevent ordinary maintenance and repair, where this does not alter or modify the historic character of the structure.

The Historic Preservation Commission must adopt design guidelines that apply to the exterior features of structures in all Historic Preservation districts. The guidelines are intended to offer assistance to property owners when building or modifying structures in the district, as well as to establish a set of standards to be used in reviewing proposals for certificates of appropriateness.

Demolition of Historic Preservation properties is strictly regulated, and while the intent of the ordinance is to preserve historic properties, the City Council recognises that

> there can be circumstances beyond the control of the owner, or situations involving public health, safety, and welfare which may result in the necessary demolition of a structure within an HP District. These situations include a building which constitutes a danger to the public health, safety and welfare, or which involves a resource whose loss does not diminish or adversely affect the integrity of the district, or which imposes an economic hardship on its owners.[235]

4.3.4 Types of Historic Property Affected

A) Historic Districts and Properties Designated in Phoenix

Table 2 shows the Historic Districts that have been designated in Phoenix, and the date of designation. There are also 181 individual properties in the Phoenix Register designated with Historic Preservation Zoning, including such buildings as Carnegie Public Library, the Bobby Brown café, and various warehouses dating from the 1920s and 1930s.

B) The Villa Verde Historic District

A good example of the criteria used to designate a Historic Preservation Districts, and the type of architectural value and special interest that have to be present, is the Villa Verde District. The example used shows how similar the concept of Historic District in a local ordinance is to the Conservation Area used in United Kingdom law.

Designation of the Villa Verde Historic Preservation District was resident led, with research being carried out by private individuals and being brought to the attention of the Historic Preservation

structure, site or district and shall include but not be limited to elements of size, scale, massing, proportions, orientation, surface textures and patterns, details and embellishments and the relation of these elements to one another and 2. Conformance with the guidelines approved by the HP Commission.

235 Historic Preservation Ordinance City of Phoenix 813.

Commission. In 1996 the district was given priority for future designation and a consultation process was initiated.

Table 2: Historic Districts designated in Phoenix.

Alvarado, declared November 1992 (Period of Significance: 1907–1933) *Ashland Place declared May 1992; boundary expansion January 2003 (Period of Significance: 1920–1940)* *Brentwood declared April 2003 (Period of Significance: 1926–1956)* *Campus Vista declared April 2003 (Period of Significance: 1939–1956)* *Cheery Lynn declared February 1994 (Period of Significance: 1928–1945)* *Coronado declared November 1986; boundary adjustment January 2000 (Period of Significance: 1907–1942)* *Country Club Park declared January 1993 (Period of Significance: 1939–1946)* *Del Norte Place declared July 1993 (Period of Significance: 1927–1945)* *Earll Place declared April 2003 (Period of Significance: 1927–1942)* *East Alvarado declared May 1992 (Period of Significance: 1929–1942)* *East Evergreen declared November 1999 (Period of Significance: 1909–1929)* *Encanto Manor declared February 2006 (Period of Significance: 1945–1959)* *Encanto Vista declared April 2003 (Period of Significance: 1943–1953)* *Encanto-Palmcroft declared March 1987; boundary expansions June 1999, September 2005 (Period of Significance: 1920–1952)* *F.Q. Story declared March 1988; boundary adjustments March 2000, January 2004 (Period of Significance: 1921–1942)* *Fairview Place declared February 1994 (Period of Significance: 1928–1948)* *Garfield declared April 1989 (as*	*North Garfield declared June 1997 (as Diamond Street Historic District); December 2002 (as Moreland Street Historic District); district consolidation, expansion and name change February 2005 (Period of Significance: 1887–1942) Oakland declared September 1988 (Period of Significance: 1887–1951)* *Phoenix Homesteads declared January 1990 (Period of Significance: 1935–1937)* *Pierson Place declared November 2005 (Period of Significance: 1924–1956)* *Roosevelt declared September 1986 (Period of Significance: 1895–1930)* *Roosevelt Park declared March 2003 (Period of Significance 1924–1942)* *Villa Verde declared January 1999 (Period of Significance: 1928–1940)* *Willo declared July 1990; boundary expansion June 2006 (Period of Significance: 1910–1950)* *Windsor Square declared July 1996 (Period of Significance: 1912–1945)* *Woodland declared February 1989; boundary expansion December 2002 (Period of Significance: 1880–1935)* *Woodlea declared January 1999 (Period of Significance: 1928–1949)* *Yaple Park declared June 1997 (Period of Significance: 1928–1940)* **Non Residential Historic Preservation Districts** *Eastlake Park declared in June 2005 (Period of Significance: 1890–1949)* *Heritage Square – Block 14 declared in April 1989 (Period of Significance: 1895–1920)* *Matthew Henson Public Housing Project declared in June 2005 (Period of Significance: 1940–1941)*

Victoria Place Historic District); December 2001 (as Dennis Addition Historic District); December 2002 (as Garfield Place Historic District); district consolidation, expansion and name change February 2005; boundary adjustment November 2005 (Period of Significance: 1883–1942) *Idylwilde Park declared June 1991 (Period of Significance: 1928–1941)* *La Hacienda declared April 2003 (Period of Significance: 1926–1954)* *Los Olivos declared December 2003 (Period of Significance: 1906–1935)* *Margarita Place declared October 1999 (Period of Significance: 1927–1949)* *Medlock Place declared April 2003 (Period of Significance: 1926–1956)* *North Encanto declared December 2002 (Period of Significance: 1939–1950)*	*North Central Avenue Streetscape declared in July 2004 (Period of Significance: 1895–1951)* *Papago Park declared in October 1989 (Period of Significance: 1932–1946)* *Phoenix Indian School declared in October 1992; landmark designation March 2004 (Period of Significance: 1891–1931)* *Phoenix Union High School Campus declared in December 1986 (Period of Significance: 1911–1912)* *Pioneer Cemetery and Memorial Park declared in June 1991 (Period of Significance: 1880–1914)* *Pueblo Grande Museum and Archaeological Park declared in September 1991; landmark designation June 2005 (Period of Significance: AD 500–1941)* *South Mountain Park and Preserves declared in October 1989 (Period of Significance: 1933–1942)*

The Villa Verde neighbourhood is characterized by the range of styles that distinguish the Period Revival and Ranch eras of building in Phoenix, including a collection of English Cottage, Tudor Revival, Cotswold Revival, Cape Cod Revival, Spanish Colonial Revival, South-West, International, and Transitional/Early Ranch Style residences dating between 1928 and 1940. The transition from the first building era to the second becomes evident travelling north through the district, with mature palm and deciduous trees giving the neighbourhood a distinctive character. An important factor is that, of 107 houses in Villa Verde, only nine have been substantially modified, with only one being built after 1940.

The Historic significance of the district is to be considered within two contexts:

- Trends and Patterns of Residential Subdivision Development in Phoenix, 1912–1945: in the period between 1910 and 1939 Phoenix experienced rapid population growth as a result of agricultural productivity, demand for cotton, the expansion of citrus ranching, and the beginning of the Valley's tourist industry. Residential land development began a boom period after 1912, with most development being done by small builders on small lots, with houses being built to order, and very few homes built as speculation. However, by the late 1920s, demand for housing led to speculative development, and by the 1930s, larger investment companies began to appear and large-scale residential subdivisions became common. Each of these practices is manifested by the extent and scale of the various historic neighbourhoods.

- The Progression of Residential Architectural Styles, Building Materials and Construction Practices in Central Phoenix Neighbourhoods, 1912–1945: the coincidence in the 1920s and 1930s of the Spanish Colonial Revival and later Period Revival styles with the explosive residential boom during that period is possibly unique to Phoenix, and the volume of construction led to great diversity house styles. From the 1930s, budget constraints led to simpler housing styles.

The Villa Verde residential district was built almost entirely by one constructor – Frank B. Wallace. Born in New York, he came to Arizona in 1928 and began to develop the Villa Verde District. Specialized in building small, stylish homes, with a cost of around \$3000 each, the houses sold quickly and were purchased by a generally middle class customer. Wallace based his style on the fact that no two houses were to look alike, having a variety of floor plans, roof types and building materials. The houses also had distinctive colour schemes, both in the interior and exterior decorations.

The district was built outside the Metropolitan area, and part of its charm was "cooler nights" and being outside the metropolitan tax area.

Villa Verde's historical significance is, according to the Historical Commission, to be found in its representation of a

> largely intact representation of a middle-class Phoenix subdivision developed from 1928 to 1940. Flourishing at the height of the city's prosperity between World War I and World War II, the neighbourhood reflects the demand for residential development and the distinctive architecture of the period. It also represents a successful speculative venture by a single developer, providing a taste of what would follow in later years.[236]

236 The information contained here is researched from the Proposed Historic Preservation Overlay Zoning for the Villa Verde Historic District Application No. Z-78-98-4.

Chapter 3

The Protection of Historic Properties in European Union Law

1 The Illicit Import, Export and Transfer of Ownership of Cultural Property

Under European Law the protection of historic properties is considered to be the responsibility of individual member states. Each member state is free, in principle, to regulate heritage according to their own laws and values, as long as they respect European Union legislation. Nevertheless, there are certain measures that aim to protect heritage in Europe notwithstanding each individual state's own legislation.

European Union intervention in this area came about due to increased dangers of spoliation caused by the suppression of physical frontiers from the 1st of January 1993. This additional protection was felt to be necessary given that the UNESCO Convention on the Means of Prohibiting and Preventing the Illicit Import, Export and Transfer of Ownership of Cultural Property[237] was considered insufficient to provide full guarantees to protect Europe's Heritage.[238]

---

237 Paris 14th of November 1970. Entry into force 24th April 1972. The Convention was complemented by the 1995 UNIDROIT Convention on Stolen or Illegally Exported Cultural Objects. The new convention had an opt out clause which permitted member states who had previously ratified the Convention to allow Directive 93/7/EC to prevail.

238 Communication on the protection of national treasures with an artistic historical or archaeological value in the context of the removal of internal borders in 1992, COM (89) 594 final. It was deemed necessary to adopt measures at Community level as this protection could not be adequately guaranteed via the ratification of the UNESCO Convention of 1970 on the measures to be adopted to prohibit and prevent the import, export and transfer of unlawful ownership of cultural goods, or the Convention of the Council of Europe of 1985 on offences relating to cultural property.

Direct European Union intervention in heritage was at first limited to aspects relative to the export and import of movable items, and not to the full regulation of cultural heritage, and in this sense, Regulation (EEC) No 3911/92[239] on the export of cultural goods establishes a mechanism of harmonized control of the movement of items of cultural heritage.[240] The Regulation applies to cultural goods belonging to one of the categories given in its annex, and in particular the products of archaeological excavations and items from artistic, historic or religious monuments or archaeological sites which have been dismantled. The regulation establishes a system where the export of cultural goods is subject to the presentation of an export licence[241] which is valid throughout the community. It allows member states to restrict the number of customs offices competent to process exports of cultural goods and provides for mutual assistance between the administrative authorities of the member states. The regulation establishes a Cultural Goods Committee composed of representatives of the Member States who have the remit to advise the commission.

The regulation is complemented by Directive 93/7/EEC[242] which deals with the return of unlawfully removed cultural heritage, and which essentially covers cultural goods which, as they belong to the

239 Council Regulation (EEC) No 3911/92 of 9 December 1992 on the export of cultural goods Official Journal L 395, 31/12/1992 P. 0001–0005.

240 For a detailed study of this regulation, and the European Directive 93/7/EC, see HOFFMAN, B., Art and Cultural Heritage: Law, Policy and Practice, Cambridge University Press, p. 191.

241 According to Report from the Commission to the Council, the European Parliament and the Economic and Social Committee on the implementation of Council Regulation (EEC) n° 3911/92 on the export of cultural goods and Council Directive 93/7/EEC on the return of cultural objects unlawfully removed from the territory of a Member State COM/2000/0325 final "it is very clear that there are major differences between the Member States as regards the number of licences applied for and issued. This is the case in particular in the United Kingdom (a total of 38,000 licences for all categories taken together), France (with more than 8,000 licences), Italy (2,800 licences) or Germany (1,800 licences), whereas the other Member States show a more uniform pattern but have a much smaller number of Community export licences. It is probable that these statistics are biased. This is because authorisations may have been used mistakenly to cover the transfer of cultural goods to other Member States when, in fact, under Community Regulations, such licences ought to have only been used for exports to third countries. It should also be pointed out that more than two-thirds of the export licences were issued in the United Kingdom; even though this figure is not in any way a true reflection of the reality – since the data from this country includes all types of authorisations and licences and not only those issued pursuant to the Regulation – the commercial aspect plays an important role. The United Kingdom continues to be a major centre for the purchase and sale of art, attracting objects from other Member States. This contrasts with the situation in the Mediterranean countries, where virtually all of the objects exported are part of the cultural heritage".

242 Council Directive 93/7/EEC of 15 March 1993 on the return of cultural objects unlawfully removed from the territory of a Member State Official Journal L 74, 27.3.1993, p. 74–79.

categories referred to in its Annex (which are the same as those listed in the Annex to the Regulation), are classified as national treasures possessing artistic, historic or archaeological value under the terms of the legislation or administrative procedures of the Member States. As for cultural goods which do not belong in one of the categories in the Annex, only public collections and the inventories of ecclesiastical institutions are covered by the Directive.

This exception apart, national treasures which are not "cultural goods" within the meaning of the Annex are excluded from the Directive and thus governed by the national legislation of the Member States in accordance with the rules of the Treaty. Article 14 (1) of the Directive does allow that Member States may extend the obligation to return cultural objects to cover categories of objects other than those listed in the Annex. The obligation to return enshrined in the Directive applies only to cultural goods which have been unlawfully removed from the territory of a Member State. Unlawfully removed is defined as:

- removed from the territory of a Member State in breach of its rules of the protection of national treasures;
- removed from the territory of a Member State in breach of Regulation 3911/92/EEC on the export of cultural goods; or
- "not returned at the end of a period of lawful temporary removal or any breach of another condition governing such temporary removal".[243]

According to the Committee report, one of the major advantages of the regulation has been to

> draw the attention of the responsible administrations and customs authorities to the importance of the art world and, in particular, to the existence of illegal trade in cultural property. However, this interest has not really been converted into concrete results. The effectiveness of customs controls on the export of cultural goods to a third country varies according to the Member State's attitude and behaviour as regards various aspects of culture and art. It is plain, therefore, that although the Regulation has successfully standardised the formalities and documentation needed at Community level for the export of cultural goods, and that this has not been matched by any real change in behaviour and attitudes on the question of genuine protection for cultural goods at Community level. Indeed, for the time being the protection of cultural goods still concentrates on those goods belonging to the national heritage.

243 op.cit. Indent 2.2.

The Committee also felt that the regulations have had a "positive impact on the protection of cultural goods, if only by virtue of raising awareness among the various parties (public administrations, private organisations and operators involved in various capacities in the fields of culture and art) and a growing realisation of the importance of protecting national cultural heritage". The commission also points out that

> an awareness of the existence – and the need for protection and defence – of goods which are part of the cultural heritage of other Member States is beginning to take shape, even if priority is given to the protection of the national heritage at the expense of the rest. This realisation has been translated in practice by practical measures such as the adoption of specific laws or rules, the publication of explanatory brochures for the general public or the setting up of systems designed to draw attention to the implications of protecting cultural heritage and to the need for greater vigilance, particularly with regard to the issuing of Community export licences. In this context, and among the protection measures put in place, the competent authorities of most Member States, before issuing an export licence for a particular object, exercise a pre-emptive right (preferential right of purchase by the State itself) or make prior enquiries on the domestic market – with museums or other Institutions. These procedures have led in certain cases to the export licence application being rejected.[244]

2 Directive 85/337/EEC Regulating Environmental Impact Statements and the Protection of Historic Buildings

The regulation of Environmental Impact Statements for individual projects and for long term plans has been the European Community's most important contribution to protection of historic properties. There are two Directives that regulate environmental impact statements: Directive 85/337/EEC[245] on the assessment of the effects of certain public and private projects on the environment (amended substantially by Council Directive 97/11/EC)[246] and Directive 2003/35/EC.[247] The Directive on the assessment of the effects of certain public and private projects on the environment, commonly known as the Environmental Impact Statement Directive, provides for the assessment of the

244 COM/2000/0325 final. Indent 3.2 p. 7.

245 Official Journal No L 175 , 05/07/1985 P. 0040–0048.

246 Official Journal No. L 073 , 14/03/1997 P. 0005.

247 Directive 2003/35/EC providing for public participation in respect of the drawing up of certain plans and programmes relating to the environment and amending with regard to public participation and access to justice Council Directives 85/337/EEC and 96/61/EC. Official Journal 156 25th June 2003.

environmental effects of those public and private projects which are likely to have significant effects on the environment.

According to this Directive, projects likely to have significant effects on the environment by virtue of their nature, size or location are made subject to a requirement for development consent and an assessment with regard to their effects on the environment. This “environmental impact assessment” can either be integrated into existing procedures for development consent, or into other relevant procedures, or procedures specially developed to implement this Directive. The Directive specifically allows the incorporation of the Environmental Impact Assessment into the procedure developed for the integrated pollution prevention and control (IPPC) Directive.

The Directive uses the legal technique of an annex to include all those projects that are to be subject to Environmental Impact Statements. Those included in Annex 1, which can be defined as major public works projects, should always be subject to an Impact Statement, while those in Annex 2 are to be studied by the member states on a case by case basis.[248] When deciding on a case-by-case basis whether an EIS should be carried out, the Directive gives an indication on the criteria to be used.

The decision must be based on:

- Characteristics such as the size of the project, the use of natural resources, and pollution and nuisances;
- The location of the project, especially with regard to the environmental sensitivity of geographical areas likely to be affected by the project, taking into especial account the existing land use, and the absorption capacity of the natural environment, with one of the factors being the effect on landscapes of historical, cultural or archaeological significance;
- The characteristics of the potential impact paying special attention to the magnitude and complexity of the impact, the probability of the impact, and the duration, frequency and reversibility of the impact.

248 Annex 1 has 21 categories of project, including, amongst others, the construction of lines for long-distance railway traffic and of airports with a basic runway length of 2100 m or more, the construction of motorways and express roads, crude oil refineries and nuclear power stations. Annex 2 includes, amongst others, the manufacture of ceramic products, in particular roofing tiles, bricks, refractory bricks, tiles, stoneware or porcelain, the manufacture of vegetable and animal oils and fats, projects for the use of uncultivated land or semi-natural areas for intensive agricultural purposes and the construction of industrial installations for the production of electricity, steam and hot water.

The Environmental Impact Assessment has to include the following elements, outlined in Annex IV of the Directive:

- A description of the project, including in particular a description of the physical characteristics of the whole project and the land-use requirements during the construction and operational phases, a description of the main characteristics of the production processes, an estimate, of expected residues and emissions;
- An outline of the main alternatives studied by the developer and an indication of the main reasons for this choice, taking into account the environmental effects;
- A description of the aspects of the environment likely to be significantly affected by the proposed project, including the architectural and archaeological heritage and landscape;
- A description of the likely significant effects of the proposed project on the environment resulting from the existence of the project, the use of natural resources, the emission of pollutants, the creation of nuisances and the elimination of waste, and the description by the developer of the forecasting methods used to assess the effects on the environment;
- A description of the measures envisaged to prevent, reduce and where possible offset any significant adverse effects on the environment.

The Directive also provides for extensive public participation in the process. Both the request for development consent and the EIS should be made available to the public within a reasonable time, in order to give the public concerned the opportunity to express an opinion before the development consent is granted. The results of these consultations have to be taken into account by the relevant authorities before a decision on whether to grant development permission, and on what conditions are to be attached to the permit.

In the two European Member States object of this study, the implementation of the Directive has been different. In the United Kingdom, the planning permission legislation already covered most of the previsions of the Directive, and only relatively minor modifications to legislation were required. In Spain however, important, specific legislation has been passed at a national and regional level to ensure compliance.

2.1 The Application of Directive 85/337/EEC in Spanish Law

European Directives on Environmental Impact Assessment have been incorporated into Spanish law through the following statutory instruments: Real Decreto Legislativo 1302/1986, of the 28th of June[249] on Environmental Impact Assessment, substantially modified substantially to incorporate new European Directives by Law 6/2001,[250] and Law 27/2006[251] which introduced important modifications on public involvement in Environmental Impact Assessments. Finally, in January 2008 Real Decreto Legislativo 1/2008, of the 11th of January consolidates all previous modifications into one single legal text.

RDL 1/2008 in its article 1 establishes as the principal object of the law the establishment of a legal regime applicable to Environmental Impact Assessment of projects that consist of the carrying out of works, installation or any other of the activities enumerated in Annexes 1 and 2, which include the same activities as those in the European Directives. The law aims to insure the integration of environmental considerations into the relevant projects by incorporating the Environmental Impact Assessment into the planning permission process which already exists. The environmental assessment should, as indicated by European Legislation, include considerations of the effect of the project on historic properties.[252]

Spanish legislation opts to make Environmental Impact Assessment obligatory for all projects included in Annex 1, making the assessment for those projects included in Annex 2 subject to a discretionary decision by the public authority, based on the criteria included in Annex 3. The decision to exclude a project from an Environmental Impact Assessment must be motivated and made public. Also included as obligatory are those projects not included in Annex 1 which may affect Nature Net 2000 sites.[253]

Before a project is exempted from the procedure, a solicitude must be presented to the environmental authority containing a description of

249 BOE nº 155, 30 de junio de 1986.

250 BOE n. 111 de 9/5/2001.

251 BOE núm. 171/2006, de 19 de Julio de 2006.

252 REAL DECRETO LEGISLATIVO 1/2008, de 11 de enero, por el que se aprueba el texto refundido de la Ley de Evaluación de Impacto Ambiental de proyectos, BOE núm. 23 Sábado 26 enero 2008 p. 4986, Article 1.

253 A European Union wide network of nature protection areas established under the 1992 Habitats Directive.

the project, possible alternatives, expected environmental effects and measures to mitigate them.

The law establishes as authorizing administrations the Environment Ministry in those cases where projects are subject to government authorization, and the relevant regional environment agencies where the authorization is made at a regional level.

The procedure to be followed includes the following steps: A scoping report is prepared and presented with the initial request for the carrying out of an Environmental Impact Assessment. The scoping report should include at least the following: a report on the characteristics and geographical location of the project; a broad outline of the principal alternatives to the project and the potential impact of each of them and a brief description of the area and elements of the environment affected by the project. The scoping report is then passed on to relevant public bodies, including those with responsibilities for cultural heritage. The public authority may also consult at this initial stage with other public or private bodies that may have an interest in the project. The public authority, within the time period of three months must inform the developer of the necessary scope of the Environmental Impact Assessment.

Having received the assessment of the public authority of the necessary scope of the Environmental Impact Assessment, the developer must present a full Environmental Impact Assessment, which should include, at least the following:

- a general description of the project, including estimates of the amount of waste and energy use;
- a description of the main alternatives studied and a full justification of the main reasons behind the final choice, taking into account environmental considerations;
- an assessment of the main effects of the project on, amongst other things, the historic, archaeological and architectural heritage;
- measures proposed to mitigate or eliminate those negative effects;
- an ongoing programme of environmental reporting;
- a brief presentation of the study and its main conclusions in ordinary, understandable language.

The public authority has a responsibility to make all available information and documentation available to the developer in order to assist in the preparation of the assessment, which should include a study of both direct and indirect negative impacts.[254] The impact on the environment can be presented as major, minor, negative, positive, direct, indirect or secondary. A negative effect includes as "that which produces a loss in natural, cultural landscape...or the character and personality of a determined locality".

The law requires public involvement in the procedure, obliging the administration to make information available on all stages of the process to the public in general, and specifically, to those organisations that are involved in the protection of elements of the environment that may be affected. This prevision ensures the involvement of organisations dedicated to the protection of cultural heritage in the area affected.

The next stage in the process is the Declaration of Environmental Impact, in which the responsible environment agency issues a final document laying out the principal environmental effects of the project and the mitigating measures that must be incorporated before planning permission is granted.

2.2 The Application of Directive 85/337/EEC in United Kingdom Law

The first Environmental Impact Assessment Directive was incorporated into Law with the Town and Country Planning (Assessment of Environmental Effects) Regulations 1988. The regulations imposed

---

254 Real Decreto 1131/1988 Annex 1.

The types of impact included in the Annex are:

- simple impact: an impact that only appears over one environmental indicator;
- cumulative impact: prolonged, with a progressive increase in its level of impact synergic impact. The impact is increased by the accumulation of different factors;
- short medium or long term impact: an impact that can be measured, respectively, within a time period of one year, five years, or greater permanent or temporary impact;
- reversible or irreversible impact: the impact can be assimilated or reversed, or not, as the case may be, by the natural capacity of the environment to recover;
- recoverable or unrecoverable impact: the environment can, or cannot be recovered by natural or human action;
- periodic impact, irregular, continuous or non-continuous impact;
- critical impact: an impact above an acceptable level. A permanent loss of environmental quality will be produced even after mitigating the impact;
- severe impact: recovery of the environment requires mitigation and protective measures;
- moderate impact: recovery of the environment does not require mitigating measures;
- compatible impact: the effect on the environment will last only as long as the project and will cease upon termination.

new procedural requirements in connection with the consideration of applications for planning permission under Part III of the Town and Country Planning Act 1971 and the deemed grant under the Town and Country Planning General Regulations 1976 of planning permission for development by local planning authorities or for development on land they own.

The projects affected by the regulations are included in Schedule 1 of the regulations (projects that are in most cases subject to the preparation of an Environmental Report) and in Schedule 2 (projects that are only affected if it would be likely to have significant effects on the environment by virtue of factors such as its nature, size and location).

The content of the Environmental Statement is regulated in Schedule 3 of the regulations, which reproduces Annex 1 of the European Directive. The process includes extensive requirements for public information and involvement in the process. Regulation 13 requires that the developer seeking planning permission must publish a notice in a local newspaper with full information about the project and where documentation is available.[255]

There are also separate regulations for Environmental Assessment in specific areas. For example the Environmental Impact Assessment (Uncultivated Land and Semi-natural Areas) (England) Regulations 2001 regulate the Directive for projects for the use of uncultivated land and semi-natural areas in England for intensive agricultural purposes, and the Water Resources (Environmental Impact Assessment) (England and Wales) Regulations 2003 regulate the Directives for water management projects for agriculture in England and Wales.

---

255 The notice must contain at least the following:

- name of the project applicant stating that he is the applicant for planning permission and the name and address of the local planning authority;
- the date on which the application was made and, if it be the case, that it has been referred to the Secretary of State for determination or is the subject of an appeal to him;
- the address or location and the nature of the proposed development;
- that a copy of the application and of the plans and other documents submitted with it together with a copy of The environmental statement may be inspected by members of the public at all reasonable hours;
- an address in the locality at which those documents may be inspected, and the latest date on which they will be available for inspection;
- that copies may be obtained there so long as stocks last. If a charge is to be made for a copy, the amount of the charge;
- that any person wishing to make representations about the application should make them in writing, to the local planning authority or (in the case of an application referred to the Secretary of State or an appeal) to the Secretary of State.

3 Directive 2001/42/EC on the Assessment of Strategic Plans and the Protection of Historic Buildings

The Strategic Environmental Assessment Directive (SEA Directive)[256] has its origins in the Protocol on Strategic Environmental Assessment to the convention on environmental impact assessment in a transboundary context. The Directive aims to

> provide for a high level of protection of the environment and to contribute to the integration of environmental considerations into the preparation and adoption of plans and programmes with a view to promoting sustainable development, by ensuring that, in accordance with this Directive, an environmental assessment is carried out of certain plans and programmes which are likely to have significant effects on the environment.

The Directive determines that plans which are prepared for agriculture, forestry, fisheries, energy, industry, transport, waste management, water management, telecommunications, tourism, town and country planning or land use and which set the framework for future development consent of projects subject to EIS as determined by the Environmental Impact Directive are subject to the preparation of an Environmental Assessment. Those plans and programmes that determine the use of small areas at local level and minor modifications to plans and programmes are not subject to this Directive except where member states determine that they are likely to have significant environmental effects.

The provisions of the Directives are similar to the Environmental Impact Directive, insofar as the content of the Impact Report, which must contain, amongst other provisions, the following:

- an outline of the contents, main objectives of the plan or programme and relationship with other relevant plans and programmes;
- the relevant aspects of the current state of the environment and the likely evolution thereof without implementation of the plan or programme;
- the environmental characteristics of areas likely to be significantly affected;
- the environmental protection objectives, established at international, Community or Member State level, which are

256 Official Journal of the European Communities. 21.7.2001.

relevant to the plan or programme and the way those objectives and any environmental considerations have been taken into account during its preparation;

- the likely significant effects on the environment, including cultural heritage, architectural and archaeological heritage and landscape;
- the measures envisaged to prevent, reduce and as fully as possible offset any significant adverse effects on the environment of implementing the plan or programme;
- an outline of the reasons for selecting the alternatives dealt with, and a description of how the assessment was undertaken including any difficulties (such as technical deficiencies or lack of know-how) encountered in compiling the required information;
- a non-technical summary of the information provided.

Again, there are previsions for public consultation and the need for authorities to take into account that consultation when taking a final decision on the plan.

As far as our study is concerned, the importance of this legal provision is the obligation to prepare an Impact Statement that takes into account the effect on historic buildings and areas in local and regional plans, especially in town planning. This means that the relevant local authority, when preparing planning measures for town and country planning, tourism and land use, must factor into them any possible effects that this planning may have on architectural and cultural heritage.

Again, the implementation of the SEA Directive has been different in the two countries studied in this publication.

3.1 The Application of Directive 2001/42/EC in Spanish Law

Directive 42/2001 (the SEA Directive) is incorporated into Spanish Law by Law 9/2006.[257] The law, which includes amongst its principle objectives, "fomenting sustainable development, achieving a high level of protection of the environment and contributing to the integration of environmental concerns in long term plans and programmes by the carrying out of Environmental Impact Statements". Once again it is worth underlining that the importance of this law to our study is that it

257 BOE número 102 29/4/2006.

includes “cultural heritage” in its definition of environment, ensuring that plans and programmes produced by public administrations factor in any negative impacts they may have on historical buildings.

The law defines plans and programmes as “those strategies, directrices and proposals made by a public administrative agency developed to satisfy social needs that are not directly executable, but are to be developed through a multitude of different projects”.

In order to be covered by the law, the plan or programme must be liable to have a significant impact on the environment, including cultural heritage. All plans and programmes that establish a framework for projects that may be subject to Environmental Assessment in the areas of agriculture, aquaculture, fishing, energy, mining, industry, transport, waste disposal, water resources, use of the maritime public domain, telecommunications, transport, tourism, town or territorial planning or land use. The plans must also be elaborated or approved by an administrative agency, and must be obligatory by law or be elaborated upon request by the government.

Where the plan or programme applies to a reduced geographical area, or where it is a minor alteration of an existing plan, it is only subject to the law where they may have significant effects on the environment, including cultural heritage. The subjection of these plans to Environmental Impact Assessment is decided on a case-by-case basis after consulting with relevant agencies, including agencies with responsibility for the protection of cultural heritage and historic buildings.

The law establishes that the relevant planning law must introduce procedures to include in the relevant administrative process an Environmental Impact Assessment which must include:

- the presentation of an Environmental Report, whose level of detail and amplitude must be determined by the environment agency;
- a period of public audience and consultation;
- the taking into consideration of the previous phases in the final decision on approval of the plan; and
- procedures to ensure the availability of all information on the decision making process to the general public.

The Environmental Report must include all elements included in Annex 1 of the law, which reproduces Annex 1 of the SEA Directive, previously discussed. The law establishes a “scoping” process, where the extent of study to be included in the Environmental Report is consulted with relevant agencies and bodies, including necessarily those involved with cultural heritage. Once again the law allows private organisations that have an interest in heritage matters to become involved in the process.

3.2 The Application of Directive 2001/42/EC in United Kingdom Law

In the United Kingdom, as Therivel points out, assessment of planning before the SEA Directive was

> rapid, subjective and in house. Over time it had evolved from a mostly environmental focus to also take on board social and economic issues; from a stand-alone environmental test to an approach that aimed to integrate sustainability and plan making, and from a minimal approach into a slightly broader process that sometimes included public participation, description of the baseline environment and consideration of sustainable alternatives to the plan.[258]

Appraisal of development plans was “recommended” by the government, with a three stage process; identifying environmental components such as air quality and urban liveability that could have been affected by the development plans; ensuring that the plan was in accordance with government environmental and planning advice and; determining whether the plans objectives and policies were internally consistent.

The appraisal plans were normally carried out within the planning authority itself, and could involve consultation with outside groups. Again, according to Therivel, “although it implied that the plan could be changed in response to the appraisal findings, this was not a key message”.[259]

The SEA Directive has been implemented in England by the Environmental Assessment of Plans and Programmes Regulations 2004, with separate regulations being published in the same year for Scotland, Northern Ireland and Wales.

The Regulations lay out that for those plans and programmes that are subject to preparation or adoption by an authority at national, regional or local level, or are prepared by an authority for adoption, through a legislative procedure by Parliament or Government; and, in either case, are required by legislative, regulatory or administrative

---

258 THERIVEL, R. Strategic Environmental Assessment in Action, Earthscan 2004, p. 22.
259 THERIVEL, R. op. Cit, p. 22.

provisions, and whose first formal preparatory act of a plan or programme was on or after 21st July 2004, the responsible authority shall carry out, or secure the carrying out of, an environmental assessment, during the preparation of that plan or programme and before its adoption or submission to the legislative procedure.[260]

To be subject to these regulations, the plan must be prepared for agriculture, forestry, fisheries, energy, industry, transport, waste management, water management, telecommunications, tourism, town and country planning or land use, and must set the framework for future development consent of projects listed in the Annexes of the Environmental Impact Directive.

Again, the plan or programme may be exempted from the preparation of a Strategic Assessment if it determines the use of a small area at local level or is a minor modification to a plan or programme. The exemption of a plan or programme from the Assessment procedure can only be authorised after consultation with the four consultation bodies: the Countryside Agency, the Historic Buildings and Monuments Commission for England (English Heritage), English Nature and the Environment Agency.

The Assessment should assess the effects on the environment – including historical and architectural heritage – of implementing the plan or programme and propose reasonable alternatives taking into account the objectives and the geographical scope of the plan or programme.[261]

---

260 The Environmental Assessment of Plans and Programmes Regulations 2004 (Statutory Instrument 2004 No. 1633).

261 The report must contain, as per schedule 2 of the regulations, the following information:

1. An outline of the contents and main objectives of the plan or programme, and of its relationship with other relevant plans and programmes;
2. The relevant aspects of the current state of the environment and the likely evolution thereof without implementation of the plan or programme;
3. The environmental characteristics of areas likely to be significantly affected;
4. Any existing environmental problems which are relevant to the plan or programme including, in particular, those relating to any areas of a particular environmental importance, such as areas designated pursuant to Council Directive 79/409/EEC on the conservation of wild birds and the Habitats Directive;
5. The environmental protection objectives, established at international, Community or Member State level, which are relevant to the plan or programme and the way those objectives and any environmental considerations have been taken into account during its preparation;
6. The likely significant effects on the environment, including short, medium and long-term effects, permanent and temporary effects, positive and negative effects, and secondary, cumulative and synergistic effects, on issues such as: a) biodiversity; b) population; c) human health; d) fauna; e) flora; f) soil; g) water; h) air; i) climatic factors; j) material assets; k) cultural heritage, including architectural and archaeological heritage; l)

The regulations also include previsions for public consultation, and a statutory duty to continually monitor the significant environmental effects of the implementation of each plan or programme with the purpose of identifying unforeseen adverse effects at an early stage and being able to undertake appropriate remedial action.

The SEA assessment and the Draft plan must be made available for public consultation at the same time. Following the consultation process, a statement must be made indicating how the comments and suggestions received have affected the plan.

In summary, the SEA process forces United Kingdom planning authorities to assess whether their plans and programmes are subject to the SEA process, and if they are, to carry out a major assessment of the effect of the plans on, amongst other things, historical heritage in the affected area. This process must be carried out in consultation with English Heritage and "public consultees" (persons who, in the authority's opinion, are affected or likely to be affected by, or have an interest in the decisions involved in the assessment and adoption of the plan or programme concerned). The process must identify sites within the area of the plan, and assess the effects on them of its provisions, explaining and incorporating measures to mitigate these effects wherever possible.

4 European Community Projects for the Protection of Cultural Heritage as a Tourist Attraction

The European Union has also been involved in the protection of cultural heritage through the funding of various community wide projects, especially the ISTECH,[262] SAVE ART[263] and ARCHEO[264]

landscape; and m) the inter-relationship between the issues referred to in sub-paragraphs a) to l);

7. The measures envisaged to prevent, reduce and as fully as possible offset any significant adverse effects on the environment of implementing the plan or programme;
8. An outline of the reasons for selecting the alternatives dealt with, and a description of how the assessment was undertaken including any difficulties (such as technical deficiencies or lack of know-how) encountered in compiling the required information;
9. A description of the measures envisaged concerning monitoring;
10. A non-technical summary of the information provided under paragraphs 1 to 9.

262 The ISTECH project is aimed at developing new ways of reinforcing monuments located in areas prone to seismic activity.

263 The SAVE ART project deals with a new, 'green' conservation method to preserve old books, textiles and antiquities in an easy, low-cost way. Instead of using toxic chemical compounds that have short and long-term hazards for people, this new tool reduces the amount of oxygen to a minimum level, which kills the insect pests. This new portable tool is also appropriate for domestic works of art, as the users can not only be large libraries or museums but also

projects. There is also the YOUTH programme, which finances youth exchanges with a cultural theme, for example voluntary work by young people to restore the cultural heritage and youth initiatives in the arts.

Heritage is also seen within the community as an important element in the promotion of sustainable tourism. The Commission report "Basic orientations for the sustainability of European tourism"[265] points out that "the sustainability of tourism concerns particularly important aspects such as competitive and socially-responsible tourism enterprises, the possibility for all citizens to participate in tourism, good employment opportunities in the sector and benefits from tourism activities for local communities. It requires the preservation of cultural integrity, the incorporation of environmental issues and of the protection of heritage resources in tourism measures".[266] Heritage resources are therefore seen as a basic resource of a tourist destination.[267] According to the commission,

> ensuring good prospects for European tourism requires the preservation of its regional diversity as a competitiveness asset, ensuring quality and considering community well-being in tourist destinations, conserving or restoring European coastal zones and mountain areas, protected areas and cultural heritage sites as the basis for attractive destinations, protecting the environment and heritage resources, and using them wisely, and promoting sustainable inter and intra-destination mobility. It also depends on a level playing field for the local providers in relation to those who market the tourist product and requires tourism-generated profits to flow back to the destinations.[268]

In conclusion, European policy on the protection of historic property has moved from the protection of movable cultural heritage against spoliation and illegal exportation to include important legislation including historic properties in those items that must be considered when preparing Environmental Impact Statements. The Commission has also focused on the promotion of cultural heritage as an element of sustainable tourism, and a means of promoting cultural interchange within the EU.[269]

---

private collection owners. Two Italian SMEs, together with research centres in Spain, Italy, and the UK, and a Swedish Museum have developed the project jointly.

264 The ARCHEO project is proving to what extent 'climate change' and 'greenhouse effects' occurred in Venice before this century. By studying Canaletto's famous – and extremely accurate – paintings, researchers were able to reconstruct the Venice's climate and changing sea levels over the years. The project was developed by research institutions in France, Italy and the UK.

265 SEC(2003) 1295.

266 SEC(2003) 1295 p. 5.

267 SEC(2003) 1295 p. 7.

268 SEC(2003) 1295 p. 7.

269 See especially the Culture 2007 funding programme for the European Union.

Chapter 4

The Protection of Historic Properties in United Kingdom Law

In the United Kingdom, legislation to protect historic properties is centred in three main areas: the protection of monuments through the designation of scheduled monuments, the protection of individual buildings worthy of special attention through a listing system, and legislation that protects particular areas, through conservation area designation. The result is a complex system with great power given to local authorities in the designation and protection of historic buildings, through which a comparatively large number of buildings are subject to protection, especially through the listing system and the designation of conservation areas. Our study deals principally with the law as it applies to England, without entering into details on the differences that apply in Wales, Northern Ireland or Scotland.

1 Scheduled Monuments

National Monuments are regulated in the Ancient Monuments and Archaeological Areas Act of 1979, as amended by the National Heritage Act of 1983.

A Monument can be understood, for the purposes of the Act, as:

a) any building, Structure or work, whether above or below the surface of the land, and any cave or excavation;

b) any site comprising the remains of any such building, structure or work or of any cave or excavation; and

c) any site comprising, or comprising the remains of, any vehicle, vessel, aircraft or other movable structure or part thereof which neither constitutes nor forms part of any work which is a monument. Any machinery attached to a monument is regarded as part of the monument if it cannot be detached without being dismantled.[270]

Ancient Monuments are defined as "a) any scheduled monument; and b) any other monument which in the opinion of the Secretary of State is of public interest by reason of the historic, architectural, traditional, artistic or archaeological interest attaching to it".[271]

The Secretary of State[272] is empowered by the Act to compile a schedule of monuments, which must include all those monuments already listed under the Ancient Monuments Consolidation and Amendment Act of 1913, or any monument which has had notice served according to the Ancient Monuments Act 1931.

The Secretary of State may then include, either when compiling the list, or at any time thereafter, any monument which appears to him to be of national importance.[273] The non-statutory criteria for scheduling a monument, reproduced here from Planning Policy Guidance 16[274] are:

a) **Period**: all types of monuments that characterise a category or period should be considered for preservation.
b) **Rarity**: there are some monument categories which in certain periods are so scarce that all surviving examples which still retain some archaeological potential should be preserved. In general, however, a selection must be made which portrays the typical and commonplace as well as the rare. This process should take account of all aspects of the distribution of a particular class of monument, both in a national and a regional context.
c) **Documentation**: the significance of a monument may be enhanced by the existence of records of previous investigation or, in the case of more recent monuments, by the supporting evidence of contemporary written records.

---

270 Ancient Monuments and Archaeological Areas Act 1979 (Chapter 46) Part II (supplemental) Section 61.7.

271 Ancient Monuments and Archaeological Areas Act 1979 Part II Section 61.12.

272 At present the authority for scheduling monuments rests with the Department of Culture, Media and Sport, advised by National Heritage.

273 Ancient Monuments and Archaeological Areas Act 1979 Part I Section 1 (2) (a).

274 Planning Policy Guidance 16: Archaeology and planning. Annex II 1990.

d) **Group Value**: the value of a single monument (such as a field system) may be greatly enhanced by its association with related contemporary monuments (such as a settlement and cemetery) or with monuments of different periods. In some cases, it is preferable to protect the complete group of monuments, including associated and adjacent land, rather than to protect isolated monuments within the group.

e) **Survival/Condition**: the survival of a monument's archaeological potential both above and below ground is a particularly important consideration and should be assessed in relation to its present condition and surviving features.

f) **Fragility/Vulnerability**: highly important archaeological evidence from some field monuments can be destroyed by a single ploughing or unsympathetic treatment; vulnerable monuments of this nature would particularly benefit from the statutory protection which scheduling confers. There are also existing standing structures of particular form or complexity whose value can again be severely reduced by neglect or careless treatment and which are similarly well suited by scheduled monument protection, even if these structures are already listed historic buildings.

g) **Diversity**: some monuments may be selected for scheduling because they possess a combination of high quality features, others because of a single important attribute.

h) **Potential**: on occasion, the nature of the evidence cannot be specified precisely but it may still be possible to document reasons anticipating its existence and importance and so to demonstrate the justification for scheduling. This is usually confined to sites rather than upstanding monuments.

The schedule currently includes about 19,700 entries, ranging from prehistoric standing stones and burial mounds to more recent structures such as collieries and wartime pill-boxes. It is important to note that buildings that are liable to be still in use are normally protected by listing rather than scheduling. There has been overlap in the past between listing and scheduling but, as Mynors points out, this overlap should gradually diminish in accordance with the government's intent to afford each type of building the specific type of protection required.[275]

1.1 The Process for Scheduling Monuments and its effects

Immediately after scheduling a monument, or modifying its status, the Secretary of State has to inform the owner and (if the owner is not the occupier) the occupier of the monument, and any local authority in whose area the monument is situated, of the action taken.[276]

Once a monument is scheduled, it becomes an offence to carry out any works resulting in the demolition or destruction of or any damage to the monument, any works for the purpose of removing or repairing a scheduled monument or any part of it or of making any alterations or additions thereto; and any flooding or tipping operations on land in, on or under which there is a scheduled monument, without receiving the specific consent of the Secretary of State,[277] in the form of a Scheduled Monument Consent.

The Scheduled Monument Consent order may be unconditional, or may include conditions including that "a person authorized by the Secretary of State be afforded an opportunity, before any works to which the consent relates are begun, to examine the monument and its site and carry out such excavations therein as appear to the Secretary of State to be desirable for the purpose of archaeological investigation".[278]

The following statutory defences are available for offences related to work being carried out on a scheduled monument:

- in the case where a condition imposed in the consent order is not adhered to, a defendant can allege that he took all reasonable precautions and exercised all due diligence to avoid contravening the condition;
- in the case that a scheduled monument is damaged or demolished, a defendant can allege that that he took all reasonable precautions and exercised all due diligence to avoid or prevent damage to the monument;
- in the case that scheduled monument is damaged or demolished or where flooding or tipping operations are carried out on a

275 MYNORS, C. Listed Buildings, Conservation Areas and Monuments, Sweet and Maxwell, 2006 p. 166. This book can be considered as the definitive reference work on legislation relevant to historic property protection in the United Kingdom.

276 Ancient Monuments and Archaeological Areas Act 1979 Part I Section 1 (6).

277 Section 2 (2) (a).

278 Section 2 (5).

scheduled monument, a defendant may allege that he did not know and had no reason to believe that the monument was within the area affected by the works or (as the case may be) that it was a scheduled monument;

- a general statutory defence is also available based on the fact that the works were urgently necessary in the interests of safety or health and that notice in writing of the need for the works was given to the Secretary of State as soon as reasonably practicable.

Certain types of activity do not require a specific Scheduled Monument Consent. These activities are outlined in the Ancient Monuments (Class Consents) Order 1994, which grants consent under section 3 of the Ancient Monuments Act for the execution of works of any class or description specified as permitted works in Order's schedule, "subject to any condition specified in the said Schedule in relation to works of a particular class or description".[279]

The Act also allows for a wide range of circumstances in which any authorised person may enter a scheduled monument to ascertain its condition.[280]

The Act permits the compulsory purchase of any ancient monument "for the purpose of securing its preservation",[281] being applicable in this case the Acquisition of Land (Authorization Procedure) Act of

279 The Ancient Monuments (Class Consents) Order 1994 Section 2. These class consents include:

- certain types of Agricultural, Horticultural and Forestry works, of the same kind as those previously carried out lawfully in the same location and on the same spot within that location within the period of six years immediately preceding the date on which the works commence;
- works by British Coal Corporation or their employees;
- works by the British Waterways board;
- works for the repair or maintenance of machinery, where they do not involve a material alteration to a scheduled monument;
- works which are urgently necessary in the interests of safety or health provided that they are limited to the minimum measures immediately necessary and that notice is given to the Secretary of State as soon as reasonably practicable;
- works carried out by Historic Buildings and Monuments Commission for England (English Heritage);
- some types of works of Archaeological evaluation;
- some types of works for the maintenance or preservation of a scheduled monument or its amenities being works executed in accordance with the terms of a written agreement between the occupier of the monument and the Secretary of State or the Commission;
- some types of works for the preservation, maintenance or management of a scheduled monument being works executed in accordance with the terms of a written agreement under which the Secretary of State.

280 Ancient Monuments and Archaeological Areas Act 1979 Part I Section 6.

281 Ancient Monuments and Archaeological Areas Act 1979 Part I Section 10 (1).

1946. The purchase of the land may also be by agreement, and in this case the purchaser may also be the local authority, where that monument is in or in the vicinity of their area. Both the Secretary of State and the Local authority are also permitted by the act to accept the scheduled monument in donation.[282]

1.2 Types of Historic Property Protected by Scheduling

According to English Heritage, there are over 200 "classes" of monuments on the schedule, ranging from prehistoric standing stones and burial mounds, to castles, monasteries, and more recent results of human activity, such as collieries and wartime pillboxes. Following are examples of the variety of monuments that are included in the Schedule:

- **Stott Park Bobbin Mill**: Built in 1835 by John Harrison, Stott Park was a working mill until 1971, and has remained virtually unchanged. It was originally built to meet the need for bobbins and reels for the textile trades of Lancashire, but also produced many other timber items, such as pick axe handles. The site is comprised of a bobbin mill, two mill ponds, and the site of the now demolished Stott Park smithy;
- **Woodhenge Neolithic Henge in Salisbury**: originally known as Dough Cover, Woodhenge is a penannular earthwork enclosure strongly associated with Grooved Ware pottery. It was identified in 1925 after an aerial photography survey, and was seemingly built in the time of the Beaker people;
- **Stonehenge**: the world famous Stonehenge site, a circular earthwork enclosure with internal stone settings is both a scheduled monument and a World Heritage Site;
- **Rushton Triangular Lodge**: built in the years 1594–7 the Triangular Lodge is designed as an architectural expression of commitment to Roman Catholicism. Based on the symbol of the holy trinity, the equilateral triangle, it also incorporates symbols of Mass;
- **Orford Castle**: Both listed and scheduled, the oldest castle in England for which documentary evidence exists;
- **Haig Colliery in Cumbria**: the last deep coal mine to work in the Cumbrian coalfield, Haig Colliery was closed in 1986. The

282 Ancient Monuments and Archaeological Areas Act 1979 Part I Section 11 (2) and (3).

remaining buildings consist of two engine houses and a power station and the surviving steel-framed pithead winding gear and surrounding shed. Haig Colliery is both scheduled and listed;

- **Saltford Brass Mill in Avon**: Built in 1720 and working until the early 20th century. It was of great importance to the area through the 19th century, but was closed in 1928. Mostly of the 19th century buildings survive, along with important survivals of machinery and fittings. Saltford is a scheduled monument which includes a Listed Building on site;
- **Golden Hill Fort**: a military installation on the Isle of Wight, it was built in 1863. Originally built to protect against invasion, the fort was used as a training camp and depot in the two world wars. It then became an industrial estate. The camp was restored in 1984, and is a mix of listed buildings and a scheduled monument.

2 Listed Buildings

The guiding principles of historical building protection in England are based on the perception that

> in its broadest sense, the historic environment embraces all those aspects of the country that reflect the shaping hand of human history. Scarcely any part of England is untouched by the interaction between people and nature, which has taken place over thousands of years. Some of the most obvious features of this environment are historic buildings. England is exceptionally rich in these – great churches, houses, and civic buildings – but our understanding of the historic environment now encompasses a much wider range of features, and in particular stresses the relationship between individual buildings, and also the value of historic townscape and landscape as a whole.[283]

283 "Revisions to Principles of Selection FOR Listing Buildings" DEPARTMENT FOR COMMUNITIES AND LOCAL GOVERNMENT 8th March 2007. The document continues "There is growing appreciation not just of the architectural set pieces, but of many more structures, especially industrial, agricultural and other vernacular buildings that, although sometimes individually unassuming, collectively reflect some of the most distinctive and creative aspects of English history. More than this, our understanding and appreciation of the historic environment now stretches beyond buildings to the spaces and semi-natural features that people have also moulded, and which are often inseparable from the buildings themselves. For example, the pattern of roads and open spaces and the views they create within historic townscapes may be as valuable as the buildings. In the countryside, the detailed patterns of fields and farms, of hedgerows and walls, and of hamlets and villages, are among the most highly valued aspects of our environment. England is particularly rich in the designed landscapes of parks and gardens, and the built and natural features they contain:

The Planning (Listed Buildings and Conservation Areas) Act 1990 imposes a duty on the Secretary of State (now the Secretary of State for Culture, Media and Sport) to list buildings in order to facilitate the work of planning authorities.[284] The listing system in England is based on the interpretation of two statutory criteria: architectural interest and historic interest. To be of special architectural interest a building must be of importance in its architectural design, decoration or craftsmanship; special interest may also apply to nationally important examples of particular building types and techniques (e.g. buildings displaying technological innovation or virtuosity) and significant plan forms.

In the decision on whether to list a building, the extent to which a building contributes to the qualities of a group of buildings may be taken into account. This is known as "group value". An example of where group value would be considered is where buildings were for example "fine examples of planning (e.g. squares, terraces or model villages) or where there is a historical functional relationship between a group of buildings".[285]

The listing of a building can also include, and be influenced by, "the desirability of preserving, on the grounds of its architectural or historic interest, any feature of the building containing a man made object or structure fixed to the building or forming part of the land and comprised within the curtilage of the building". Although the presence of such a feature will increase the probability of a building being listed, they will only be sufficient to justify the listing of the building if they have sufficient interest to render the building of *special* interest.[286]

---

the greatest of these are as important to national, and indeed international, culture, as are our greatest buildings".

284 "For the purposes of this Act and with a view to the guidance of local planning authorities in the performance of their functions under this Act and the principal Act in relation to buildings of special architectural or historic interest, the Secretary of State shall compile lists of such buildings, or approve, with or without modifications, such lists compiled by the Historic Buildings and Monuments Commission for England (in this Act referred to as "the Commission") or by other persons or bodies of persons, and may amend any list so compiled or approved" Planning (Listed Buildings and Conservation Areas) Act 1990 (c. 9) Chapter I Section 1.

285 Revisions To Principles Of Selection For Listing Buildings 8th March 2007 op. cit. 6.10.

286 Revisions to Principles of Selection for Listing Buildings op. cit. 6.11. According to the guide, "the provision can be used for a variety of features; examples could include a finely paneled sixteenth century room, a fireplace and over-mantel that has been introduced from another building, or an elaborate plaster ceiling".

2.1 Criteria for Listing Buildings

The statutory criteria of historical or architectural interest must be sufficiently met to a level of "special interest", otherwise the listing system would be overburdened with buildings that, while undoubtedly incorporating interesting historical and architectural features, do not have sufficient interest to warrant their protection.

Buildings may be listed in the following grades:

- Grade I buildings are of exceptional interest;
- Grade II* buildings are particularly important buildings of more than special interest;
- Grade II buildings are of special interest, warranting every effort to preserve them.

A series of general principles guide the decision on whether to list a building. These general principles are supplemented by more specific "selection guides", published by English Heritage. These selection guides are intended to supplement, and not to take precedence over the guiding principles. The stronger the presence of features – the higher the grade of listing. The general guiding principles take into account age and rarity, aesthetic merits, selectivity and national interest. The state of repair of the building is not a factor to be considered in the listing of a building.

2.1.1 Age and Rarity

While the age and rarity criteria will vary according to the type of building in question, the following are general principles that apply to all types of building.

- Before 1700, all buildings that contain a significant proportion of their original fabric are listed;
- From 1700 to 1840, most buildings are listed;
- After 1840, because of the greatly increased number of buildings erected and the much larger numbers that have survived, progressively greater selection is necessary;
- Buildings of less than 30 years old are normally listed only if they are of outstanding quality and under threat.[287]

---

287 Revisions to Principles of Selection for Listing Buildings op. cit. 6.12.

2.1.2 Aesthetic Merits

According to PPG 15, external visual merits may be important in determining whether some types of buildings are to be listed, whilst in others it is irrelevant. For example, "buildings that are important for reasons of technological innovation, or as illustrating particular aspects of social or economic history, may have little external visual quality".[288]

2.1.3 Selectivity

This criterion is used where it is deemed necessary to preserve examples of a certain type of building. As a large amount of buildings of a certain type may still exist, the policy is to list only the most significant or representative examples of a type of building.

At the time of publication, Selection Guides for 20 different types of Listed Buildings have been published, covering agricultural buildings, commemorative structures, commercial buildings, culture and entertainment, vernacular houses, town houses, suburban and country houses, modern houses, educational facilities, gardens and parks, health and welfare, industrial buildings, law and government buildings, maritime and naval buildings, military buildings, places of worship, sport and recreational facilities, street furniture, transport buildings and utilities and communications. Although these guides have no legal force, they do offer insights into the principles that may be used to select buildings for listing.

2.2 Criteria for Listing of Industrial Buildings

The Industrial Buildings Selection Guide was published in March 2007. Taking into account the large numbers of surviving industrial buildings, National Heritage considers that "the greatest care needs to taken to establish the precise nature of a site's special interest prior to designation".

To this end, certain key issues are identified:

- **the wider industrial context** the selection guide considers that, more than any other type of building, industrial buildings need to be considered within the wider industrial setting and their specific position in the production process;

288 Revisions to Principles of Selection for Listing Buildings op. cit. 6.14.

- **regional factors** it is considered important to place industrial buildings within their regional context, a strong example being regional industrial specialization;
- **integrated sites** the guide considers that where the building was part of a site that contained various elements essential to understanding the buildings significance, it is unlikely to be listed where it is the only surviving building. A complete site is much more likely to be listed;
- **architecture and process** a site is unlikely to be listed if it is simply a block designed to house an industrial process, and its design does not reflect the specific function it was intended to fulfill;
- **technological innovation** an important factor in considering the significance of a building is whether it represented innovations in the industrial process it houses. The guide points out that technological significance may also reside in the building itself rather than the industrial process it housed;
- **historic interest** the survival of physical evidence of important industrial processes is an important criteria, "in some cases historical association with notable achievements may be sufficient to list: much will depend on the force of the historical claims, and the significance of the persons or products involved at the site in question".[289]
- **rebuilding and repair** if a building has been substantially rebuilt it may have an adverse effect on listing. The more the building maintains its original structure the more likely it is to be listed.

As an aid in the selection process, the selection guide divides Britain's industrial past into five major periods, examining the principal developments in each period on industry in England and how it may effect listing, looking at Historic buildings in five different time periods: the period up to 1700, 1700–1815, 1815–1914, and 1914 to the present day.

The guide points out that the historic industrial environment is the cumulative product of mankind's industrial activities from the earliest days – not just the more prominent impact of the last three hundred years – and much of this survives as archaeology. Early industrial

289 Industrial Buildings Selection Guide. Heritage Protection Department. March 2007 p. 3.

buildings were often on a domestic scale, and the typical production unit – the workshop – has largely disappeared or survives heavily disguised by later conversion. While wind and water mills (for grinding corn, fulling cloth or crushing raw materials) were once common, very few pre-1700 examples survive in anything like their original form and machinery is extremely rare.

For the period after 1700, during the industrial revolution, there were significant advances in ways of working, and the all-important emergence of the factory system: the emergence of a disciplined, systematised approach to production bringing together multiple powered processes on the same site. There was also a parallel development in transport systems. According the guide, these developments have left a relatively abundant physical legacy comprising furnaces, forges and engine houses, fairly complete textile mill complexes, docks and harbours, numerous toll houses, bridges and canal structures and pioneer industrial settlements.

The period after 1815 saw a spectacular growth in the industrial base in the United Kingdom, with the country being seen as the "workshop of the world". The guide points out the importance of the transport infrastructure in this period and the dependence of all industries on coal. Also, this was seen as "the classic period of the metal processing and metal working industries, and of the emergence of large-scale workshops. The pioneer blast furnaces, foundries and engine works of the earlier period were soon to be replicated and dwarfed in scale in hundreds of locations". There were also a large number of factories dedicated to weaving and wool production during this period.

Finally, the selection guide points out that the 20th century is not easily evaluated, with traditional industries in decline throughout the country. A large number of collieries have closed, and the textile industry, although re-emerging briefly in the early 20th century, is now almost non-existent.

Finally, the Selection guide examines the special interest of specific types of buildings. In the first category, extraction, the guide examines different types of collieries, metal mines, quarries and limekilns. The guide also examines different buildings dedicated to processes of manufacturing, looking at different types of mill (corn mills, paper mills and textile mills), workshops and buildings associated with drinks and food processing, such as breweries and food processing factories. The guide also examines criteria for listing engineering works and factories, and buildings dedicated to storage and distribution.

2.3 Effects on Property Owners of Listing Buildings

2.3.1 Obtaining Listed Building Consent

Listing does not impose immediate duties on owners over and above those that apply to all buildings, other than, as Mynors points out, the duty to obtain consent for works to them.[290]

The main effect of listing is the prohibition on either the demolition or the carrying out of any alteration or extension works "in any manner which would affect its character as a building of special architectural or historic interest, unless the works are authorized".[291]

Extension and alteration are considered as authorised where written consent for their execution has been granted by the local planning authority or the Secretary of State and where they are carried out in accordance with that consent and those conditions that are attached to the authorisation.

Demolition works are understood to be authorised where consent has been given and notice of the proposal to execute the works has been given to National Heritage. Provisions are included in the act for National Heritage to make a record of the building before its demolition, and for that reason, demolition is not understood to be finally authorised until either: National Heritage are given access to the building to record it for a period of at least one month; or where they have stated in writing that they have completed their recording of the building or that they do not wish do so. Once again, the demolition must be carried out according to any conditions that are attached to the consent order.

2.3.2 Building Preservation Notices

Section 3 of the Act allows public authorities to issue a building preservation notice in those cases where a building that is of special architectural or historic interest is in danger of demolition or of alteration "in such a way as to affect its character as a building of such interest".[292]

---

290 MYNORS, C. Listed Buildings, Conservation Areas and Monuments, Sweet and Maxwell, 2006 p. 213.

291 Planning (Listed Buildings and Conservation Areas) Act 1990 Part 1 Chapter II Section 7.

292 Planning (Listed Buildings and Conservation Areas) Act 1990 Part I Chapter I Subsection 3 (1) a.

The building preservation notice can be served either to the owner or occupier of the building, or, in cases of urgency, where the issuing authority feels it necessary, they may, instead of serving the notice on the owner and occupier of the building, affix the notice conspicuously to some object on the building.[293]

The building preservation notice should state that the building appears to be of special architectural or historic interest and that they have requested the Secretary of State to list the building. The notice comes into force immediately when it is served, and remains in force for six months from the day it is served, or last served. It ceases to be in force when the building is listed, or when the local planning authority is informed in writing that the Secretary of State has no intention of listing the building. In this case, the Local authority may not reissue a Building preservation notice on the same building in a period of 12 months.

The building preservation notice has the effect of applying the protective provisions applied to listed buildings "in relation to the building as if it were a listed building",[294] except for the provisions of section 59 on damage to listed buildings.

Section 29 lays out that where a building preservation notice ceases to have effect without the building finally being listed, any person who at the time when the notice was served had an interest in the building will be entitled to be paid compensation by the local planning authority in respect of any loss or damage directly attributable to the effect of the notice. The loss or damage can include compensation for any breach of contract caused by the necessity of discontinuing or countermanding any works to the building on account of the building preservation notice.

2.3.3 Listed Buildings Enforcement Notices

United Kingdom listed buildings law allows for a specific administrative procedure where works are carried out on a listed building without having received the necessary authorisation, or where works are carried out without respecting the conditions imposed.

Section 38 of the Listed Buildings Act allows a Local Authority to issue a Listed Building Enforcement Notice where it appears that works are being carried out on a listed building in the area, and that it

293 Planning (Listed Buildings and Conservation Areas) Act 1990 Part I Chapter I Subsection 4 (1).

294 Planning (Listed Buildings and Conservation Areas) Act 1990 Part I Chapter I Subsection 3 (5).

appears that the works being carried out for its demolition or for its alteration or extension in any manner which would affect its character as a building of special architectural or historic interest, or where works are being carried out under a listed building consent but the works are failing to comply with the conditions attached.

The listed building enforcement notice should specify the alleged contravention and specify the steps to be taken and the time period imposed for either restoring the building to its former state (in the case that restoration is not practical or is undesirable, the local authority may specify what works should be carried out to alleviate the works carried out without listed buildings consent) or for bringing the building to the state in which it would have been if the terms and conditions of any listed building consent which has been granted for the works had been complied with.

The listed building enforcement notice should specify the date on which it is to take effect, and must be served within 28 days of its issue, and at least 28 days before its effect date, to the owner or occupier, and any other person having an interest in the building. The order can be withdrawn at any time, and a new one issued with different conditions.

Where the enforcement notice specifies what works are to be carried out to alleviate damage, the notice is understood to substitute a listed building consent. The enforcement notice can be appealed on a variety of grounds,[295] and must be made before the order comes into

---

295 The grounds for appeal listed in Section 39 (1) are:

- that the building is not of special architectural or historic interest;
- that the matters alleged to constitute a contravention of section 9(1) or (2) do not involve such a contravention;
- that the contravention of that section alleged in the notice has not taken place;
- that works to the building were urgently necessary in the interests of safety or health or for the preservation of the building, that it was not practicable to secure safety or health or, as the case may be, the preservation of the building by works of repair or works for affording temporary support or shelter, and that the works carried out were limited to the minimum measures immediately necessary;
- that listed building consent ought to be granted for the works, or that any relevant condition of such consent which has been granted ought to be discharged, or different conditions substituted;
- that copies of the notice were not served as required by section 38(4);
- except in relation to such a requirement as is mentioned in section 38(2)(b) or (c), that the requirements of the notice exceed what is necessary for restoring the building to its condition before the works were carried out;
- that the period specified in the notice as the period within which any step required by the notice is to be taken falls short of what should reasonably be allowed;
- that the steps required by the notice for the purpose of restoring the character of the building to its former state would not serve that purpose;

effect. The lodging of the appeal automatically suspends the enforcement notice until it is resolved by the Secretary of State.[296] The appeal may be refused if the appellant does not submit it within the time limit, and the enforcement notice may be quashed if the Local Authority does not respond to requests for further information made by the Secretary of State.[297]

If an enforcement notice is not adhered to within the compliance period the authority may enter the land and carry out the works themselves, and recover from the owner any expenses reasonably incurred. The expenses are treated as a simple contract debt recoverable in any court of competent jurisdiction.[298]

2.3.4 Preventing Deterioration and Damage to Listed Buildings

Section 47 of the listed building act allows that, if it appears to the Secretary of State that reasonable steps are not being taken for properly preserving a listed building he may authorise the appropriate authority to acquire compulsorily under this section the building and any relevant land may compulsorily acquire them himself. Any person having an interest in the building may, within 28 days of the order being served, apply to a magistrate's court for an order staying further proceedings on the compulsory purchase order. If the court is satisfied that sufficient steps are being taken to protect the building, the court may agree to stay proceedings on the compulsorily purchase order.

The compulsory acquisition order may only be served if, at least three months previously, a repair notice has been issued by the appropriate authority or by the Secretary of State, and that it has not been withdrawn. A repair notice in England, when issued by the Secretary of State, must be consulted with the Historic Buildings Commission before being served or withdrawn.

Compensation under this section is based on the understanding that building consent would be granted for any works for the alteration or extension of the building or for the demolition of the building for the

---

- that steps required to be taken by virtue of section 38(2)(b) exceed what is necessary to alleviate the effect of the works executed to the building;
- that steps required to be taken by virtue of section 38(2) (c) exceed what is necessary to bring the building to the state in which it would have been if the terms and conditions of the listed building consent had been complied with.

296 Planning (Listed Buildings and Conservation Areas) Act 1990 Planning (Listed Buildings and Conservation Areas) Act 1990 Chapter IV Section 39 (3).

297 Planning (Listed Buildings and Conservation Areas) Act 1990 Chapter IV Section 40 (1).

298 Planning (Listed Buildings and Conservation Areas) Act 1990 Chapter IV Section 42.

purpose development not constituting new development.[299] In the case that the listed building has been left in a deliberate state of disrepair, the compulsory purchase order may include a direction for minimum compensation.

A local authority may also acquire under this section by agreement any building appearing to them to be of special architectural or historic interest. They may also purchase land "comprising or contiguous or adjacent" which is deemed necessary to preserve, afford access to or properly manage the building.

Where a local authority or joint planning board compulsorily, or by agreement, purchases a property, "they may make such arrangements as to its management, use or disposal as they consider appropriate for the purpose of its preservation".[300]

The act also includes provisions for the urgent protection of historic buildings that are unoccupied, without previously allowing the owner the possibility of carrying out those works himself. Section 54 allows public authorities to take urgent steps to preserve unoccupied buildings, including works for affording temporary support or shelter for the building. If part of a building is occupied, the urgent steps may only be taken on the unoccupied parts of the building. Notice must be given to the owner of the building at least 7 days before their commencement, and are recoverable from the owner by the executing authority. The owner can make an appeal on the amount of the expenses to the Secretary of State, alleging either:

a) that some or all of the works were unnecessary for the preservation of the building;
b) in the case of works for affording temporary support or shelter, that the temporary arrangements have continued for an unreasonable length of time;
c) that the amount specified in the notice is unreasonable; or
d) that the recovery of that amount would cause him hardship.[301]

299 Planning (Listed Buildings and Conservation Areas) Act 1990 Chapter V Section 49.
300 Planning (Listed Buildings and Conservation Areas) Act 1990 Chapter V Section 53.
301 Planning (Listed Buildings and Conservation Areas) Act 1990 Chapter V Section 55. English Heritage operates a grant programme to encourage and support local authorities in the use of their statutory powers to ensure that work urgently necessary for the preservation of a vacant or partly occupied listed building is undertaken.

2.4 Offences under the Listed Buildings Act

Section 9 of the act considers that the demolition or alteration without the required authorisation is considered an offence for which, on summary conviction the offender may be imprisoned for up to three months or fined, or, on conviction on indictment the offender may be imprisoned for up to twelve months and fined. In this case, the fine should take into account the "financial benefit which has accrued or appears likely to accrue to him in consequence of the offence".[302]

It is also considered an offence under Section 9 of the act to execute or cause to be executed any works in relation to a listed building under a listed building consent where that work fails to comply with any condition attached to the consent.

This offence requires therefore that the following facts concur:

a) the building is listed;

b) works are being carried out for its demolition, alteration or extension;

c) those works are executed or are caused to be executed by the defendant;

d) the works are authorised by the grant of a consent; and

e) The conditions attached to that consent are not complied with.[303]

The only statutory defence to these offences is laid out in section 3 of the Act. In order to apply this defence, the following conditions must be met:

302 Planning (Listed Buildings and Conservation Areas) Act 1990 Chapter V Chapter II Section 9 (5). The Best Practice Guidance on Listed Buildings Prosecutions points out that "the prosecution must consider whether the defendant has made, or is likely to make, any financial gain as a result of the works, and should draw this to the court's attention accordingly at sentencing. It is important that this evidence is not considered until after a guilty finding has been made as introducing this evidence during the trial could render the process unsafe and, thus, appealable. The most common form of financial gain in such cases will be the profit made by the defendant on the market value of the listed building as a result of the unauthorised works. Such profit may be reduced, of course, by the potential for enforcement action against future owners, but often remains a substantial sum. Again, it may be necessary to obtain expert evidence as to actual or potential financial gain from a qualified valuer. Prosecuting authorities are advised to obtain such evidence as part of the investigation process since it may have a bearing on the likely fine to be awarded and, thus, the decision to prosecute. Obtaining this evidence at an early stage also means that it can be used if an early guilty plea is entered, thus reducing the likelihood of a low penalty due to lack of evidence of financial gain". Best Practice Guidance on Listed Buildings Prosecutions December 2006 Department for Communities and Local Government: London p. 38.

303 "Best Practice Guidance on Listed Buildings Prosecutions" p. 43.

- works to the building were urgently necessary in the interests of safety or health or for the preservation of the building;
- it was not practicable to secure safety or health or, as the case may be, the preservation of the building by works of repair or works for affording temporary support or shelter;
- the works carried out were limited to the minimum measures immediately necessary; and
- notice in writing justifying in detail the carrying out of the works was given to the local planning authority as soon as reasonably practicable.

The Governments Best Practice Guidance on Listed Buildings Prosecutions points out that "the offences under sections 9 and 43 of the Planning (Listed Buildings and Conservation Areas) Act 1990 are all strict liability offences". The lack of any reference to the defendant's belief, knowledge, intentions or motive in the definition of the offence is what makes the offence one of strict liability. As a consequence, any general defense which relates to the state of mind of the defendant, such as insanity, intoxication, or mistake cannot be relevant.[304]

A further offence is laid down in section 43 of the act. This offence requires an enforcement notice to have been served, and that the compliance period has passed without the steps required for compliance having been taken, and that the defendant was or is the owner. A statutory defence exists, requiring that the defendant did everything he could be expected to do to secure that all the steps required by the notice were taken, or that he was not served with a copy of the notice and was not aware of its existence.

Section 59 of the Act also creates an offence where "if, with the intention of causing damage to a listed building, any relevant person does or permits the doing of any act which causes or is likely to result in damage to the building".

The following elements must therefore concur for this offence to have taken place:

- the building must be a listed building;
- the must be an element of intentionality. The responsible person must intend to damage the building;

304 "Best Practice Guidance on Listed Buildings Prosecutions" p. 45.

- the person must be a relevant person (they would be entitled to do or permit the doing of the act) – which in practical terms means they would be the owner or the occupier; and
- they have done or permit the doing of the act which either damages, or is likely to cause damage to the building.

2.5 Types of Listed Buildings Protected

There are an enormous number of buildings that are protected using the technique of listing. At present there are more that half a million listed buildings, of which approximately 90% are grade II. As examples of different types of listed buildings:

- **The Offices of the Mersey Docks and Harbour Board** first constructed in 1907, renowned for its symmetrical design, it has an eight-sided hall that reaches right up to the dome. Distinguished by a central domed roof it has cupolas in the corner reflecting the central dome;
- **Crossness Pumping Station** first constructed in the middle of the 18th century as a sewage pumping station, Crossness was officially opened by the Prince of Wales in 1865. It is now a Grade I listed building, and still contains the four original pumping engines;
- **Darley Abbey Mills Manufacturing Complex** a former manufacturing complex situated in Derby it forms the southern section of a larger cotton textile factory, and is now partially in use as small industrial estate. Darley Abbey Mills was built in 1792 and continuously remodelled and enlarged between 1788 and 1840. It is listed as Grade I;
- **The former Essex County Cricket Club Pavilion** first built in1886 for the Essex County Cricket Club, It remained the official club headquarters until 1934, and is now listed as a Grade II building;
- **Red Telephone Boxes** there are many examples of the traditional red telephone box that have been protected by listing. One example is the telephone box adjoining St Wilfrid's Church, described as "a cast-iron, type K6 telephone box, designed in 1935 by Sir Giles Gilbert Scott";[305]

---

305 NMR Number: SE 65 SW 1175.

- **Walthamstow Greyhound Stadium** built in the 1930s, Walthamstow Greyhound Stadium is considered an iconic art deco style greyhound stadium and is thought to be the best surviving and most architecturally interesting vintage greyhound stadium in the country.[306]

2.6 Incentives for the Preservation of Historic Buildings

Ownership of a listed building does of course restrict the actions of the owner more than in the case of ownership of non-listed building, and such restrictions, while justified by the general interest, do not give a right to any compensation. The administrative act of listing a building, a declaratory act which considers the building possesses certain attributes that make it worthy of listing, brings down on that building a legal framework that severely restricts its owner's rights, and may involve him in the outlay of important amounts of money.

The act does intend to offset this burden by allowing local authorities to contribute, by grant or by loan, towards the expenses incurred or to be incurred in the repair or maintenance of a listed building which is situate in or in the vicinity of their area. This possibility is also made extensive to buildings in their area which is not listed but which appear to them to be of architectural or historic interest.[307] The local authority under this section may also contribute towards the expenses incurred in the upkeep of any garden occupied with the building and contiguous or adjacent to it.

Loans made under this section may be interest free, and payment may also be renounced by the local authority, or renegotiated with the owner. Grants and loans made under Section 57 may include an agreement with the owners to allow the public to have access to the property or part of it at certain times specified in the agreement.

2.7 Tax Concessions and Historic Property

The regime of tax concessions in the United Kingdom has two main directions. On the one hand, it enables private owners to maintain their property while increasing public access, and on the other hand, when private owners can no longer afford the upkeep of their property, ensure that the property goes to an appropriate heritage organisation.

306 Waltham Forest Local Authority.
307 Section 57.

There are two main types of tax relief: Value Added Tax relief, which allows building work to be carried out more cheaply, and inheritance tax relief.

With regards to VAT, there is no system that gives general relief for works carried out on historic buildings, with the regime applicable depending on the individual circumstances of each individual building. In some cases there is a possibility of either zero-rating, or reduced rate VAT on certain kinds of repairs carried out on protected buildings. For example, repairs carried out on protected buildings used for residential purposes have a zero-rating of VAT on the supply of materials and service. In order to be eligible for zero rating, the work must be work that can only be done with Listed Building Consent, which has been given and received. The relief only applies to major alterations or demolition work, and not to minor repairs and maintenance.

There is also zero-rating of VAT on the sale of certain old buildings, with VAT being zero-rated on the first sale of flats which have been created by the conversion of a listed or scheduled building which has not previously used for residential purposes.

Another possible measure of tax relief is in the system of inheritance tax. Under United Kingdom tax law there may be exemptions from inheritance tax on historic properties under certain conditions.[308] The conditions imposed have two main aims: ensuring public access to the historic property, and the proper maintenance of the property. Maintenance Funds and Trust Funds[309] which have been set up to provide for the maintenance, repair or preservation of the property and for the provision of reasonable public access to it are also exempt. Gifts to certain organizations are also exempted. These organizations include National Heritage, the National Trust for places of Historic Interest or Natural Beauty, the Historic Churches Preservation Trust, and Local Authority or any Government Department.

308 See TCGA92/S258 (1)–(3) The following types of property may be exempt: Buildings of outstanding historic or architectural interest; Land which is needed for the protection of the character and amenities of buildings of outstanding historic or architectural interest along with any subsidiary buildings situated on that land; Land of outstanding scenic, historic or scientific interest along with any buildings on that land; Pictures, prints, books, manuscripts and other works of art of national, historic or artistic interest; Scientific collections or other things not yielding income which are of national, historic, artistic or scientific interest.

309 See TILEY, J., Revenue Law Hart Publishing 2005 p. 712.

2.8 Preservation Programmes

The National Lottery Act[310] determined that a certain proportion of money from the proceeds of the National Lottery should be allocated for expenditure on or connected with the national heritage.

The Heritage fund is composed of a chair and 14 trustees, appointed by the prime minister, and has a series of regional and local committees who assess applications locally. Since its inception, over £700 million has been spent on more than 3500 historic buildings projects. Included in the Heritage fund is the Townscape Heritage Initiative, which allocates funds for the regeneration of historic towns and cities across the UK, including many in deprived areas.[311]

English Heritage also has a large number of programmes offering grants for the preservation of historic buildings, including the following schemes: repair Grants for Places of Worship,[312] Cathedral Repairs,[313] War Memorials, Capacity Building for the Voluntary Sector,[314] Local Authorities – Area Partnership Funding and Local Authority Grants to Underwrite Urgent Works Notices. Other schemes worthy of note are:

- The Acquisition Grants for Local Authorities scheme. This scheme allows English Heritage to help local authorities in the serving of Repairs Notice under the Listed Buildings Act on historic buildings which have fallen into a serious and dangerous state of decay;
- The Local Authority Grants for Conservation Staff allows English Heritage in assessing and helping to create permanent specialist conservation posts in local planning authorities;
- The Historic Environment Enabling Programme. Under this programme, English Heritage commissions research and funds national initiatives;
- The Aggregates Levy Sustainability Fund (ALSF). This fund allows English Heritage to deal with the impacts of Aggregates extraction;

---

310 National Lottery etc. Act 1993 (c. 39).

311 Our Towns and Cities: The Future – Delivering an Urban Renaissance. Office of the Deputy Prime Minister November 2000.

312 Jointly funded by English Heritage and the Heritage Lottery Fund.

313 Provides funding to Church of England and Roman Catholic cathedrals which are listed grade I or II* and / or are situated within a conservation area.

314 Provides funding to national and local voluntary organisations for projects which promote the conservation, understanding and enjoyment of the historic environment.

- The English Heritage Historic Buildings, Monuments and Designed Landscapes grant scheme. This scheme is aimed at urgent repairs or other work required to prevent loss or damage to important architectural, archaeological or landscape features, in those cases where a listed building with Grade I or Grade II*, or a scheduled monument is at risk and is not eligible for other sources of public funding. This scheme is only aimed at urgent repairs.

3 Conservation Areas

3.1 The Designation of Conservation Areas

Conservation areas were first introduced as a concept in the Civic Amenities Act in 1967 and consolidated in the Listed Buildings Act. Section 69 of the Act states that Local authorities shall "from time to time determine which parts of their area are areas of special architectural or historic interest the character or appearance of which it is desirable to preserve or enhance, and...shall designate those areas as conservation areas".[315]

A conservation area is therefore an area that:

a) has special architectural or historical interest; and

b) is worthy of protection or enhancement

The designation of an area as a conservation area can also be carried out by the Secretary of State, although he must first consult the local authority. English Heritage can designate conservation areas in London, first consulting the relevant London Borough Council and obtaining the consent of the Secretary of State for National Heritage. There are at present more than 8,000 conservation areas in England.

While there is a duty to examine the local area to determine whether possible conservation areas do exist, there is no formal administrative procedure for the designation of conservation areas, and there is no obligation to enter into public consultation, although many local authorities do in fact do so.

An example of such a consultation process is that used by Bradford City Council, where the following steps are taken:

- the creation of a draft assessment, including summaries and maps;

315 Planning (Listed Buildings and Conservation Areas) Act 1990 Part II Section 69.

- a series of briefing sessions designed to inform those parish councils and groups in the affected area;
- a draft "Conservation Area Assessment" is made available to the public at public libraries, council offices and online, and a press release is made;
- inhabitants of the area affected receive in their homes a summary of the assessment document, along with a comments sheet and an invitation to attend a public workshop;
- after the public workshop is held, a period of a month is begun where members of the public may return comments sheets. Each comments sheet is individually replied to;
- a process of evaluation by the Conservation team is begun, and amendments to the Conservation Area assessment are made if appropriate;
- the amended documents are presented for final approval. When they have gone through the legal requirements, a press release is made;
- the final full document and summary are made available to the public at public libraries, council offices and online.

Once a conservation area is designated by the local authority, it must give notice to the Secretary of State and National Heritage, and information about the designation must be published in the London Gazette and in at least one newspaper circulating in the area of the local planning authority.

3.2 Property Restrictions in Conservation Areas

The act gives local authorities extra powers in relation to demolition, minor developments, and the protection of trees in areas designated as conservation areas.

3.2.1 Restrictions on demolition

In general, section 74[316] states that "a building in a conservation area shall not be demolished without the consent of the appropriate authority (in this Act referred to as 'conservation area consent')". Procedures for demolition are similar to listed building consent applications, and there is always a presumption in favour of retaining

316 Planning (Listed Buildings and Conservation Areas) Act 1990 Chapter V.

buildings. These restrictions do not apply to listed buildings, which of course are covered by other sections of the Act. Note that according to *Shimzu (UK) Ltd v Westminster,*[317] for demolition consent to be required the building must be completely demolished. Partial demolition is considered in this context to be an alteration.

3.2.2 Minor developments

The Town and Country Planning (General Permitted Development) Order 1995 (GPDO) freed many minor alterations on buildings from the need to request planning permission, thus allowing certain types of development to proceed without the need for a planning application (Permitted Development Rights), since planning permission for them is deemed to be granted. The system intends to reduce the number of planning applications for minor and uncontentious development, helping reduce the regulatory burden of the planning system.[318]

Minor developments can, however, be particularly damaging in conservation areas, and so the GPDO gives local councils and the Secretary of State the power to withdraw permitted development rights in conservation areas by a technique used as article 4 directives. The effect of withdrawing Permitted Development Rights is to force property owners to request planning permission for works that would otherwise be exempt,[319] thus allowing local authorities either to deny permission, or to grant it with conditions. A typical example would be forcing the use of local materials in building a wall, or using a certain type of paint on doors.

In those cases where planning permission is refused, the local authority may be liable to pay compensation, which may act as a deterrent for local authorities in the application of article 4 directives.[320]

Typical examples of the effects of Article 4 Directives are those associated with the Lewes Conservation Area in Sussex. This area has an Article 4 directive which requires Conservation Area Planning Permission for works that front a highway, river, watercourse or open space, where the work proposed is included on the following list:

---

317 *Shimzu (UK) Ltd v Westminster* CC [1996] 3 PLR 89.

318 REVIEW OF PERMITTED DEVELOPMENT RIGHTS: Office of the Deputy Prime Minister 2003 p. 7.

319 THOMAS, K. Development Control: Principles and Practice Routledge Press, 1997, p. 42.

320 REVIEW OF PERMITTED DEVELOPMENT RIGHTS: Office of the Deputy Prime Minister 2003 p. 269.

- any alteration to a roof, including roof coverings, roof lights and solar panels;
- building a porch;
- erecting sheds and temporary buildings;
- building a swimming pool;
- building a hard standing;
- installing a satellite dish or antenna;
- removing, building or altering chimneys;
- removing, or building walls, gates, fences or other means of enclosure;
- exterior painting of previously unpainted surfaces or changes of external colour schemes;
- any enlargement or improvement or alteration such as an extension or conservatory;
- any change to doors or windows, including changed materials, details and designs and types of decorative finish.

3.2.3 Trees in Conservation Areas

A tree preservation order (TPO) is an order made by a Local Authority to protect trees and woodland, making it an offence to cut down, uproot, willfully damage or destroy a tree without the local planning authorities' permission. This restriction normally applies only to those trees that are covered by a specific TPO. However, under Section 211 of the Town and Country Planning Act, anyone proposing to cut down or carry out work on a tree in a Conservation Area is required to give the Local Planning Authority six weeks prior notice, giving the local authority the possibility of applying a TPO to the tree. Failure of the owner to advise the local authority is an offence.[321]

3.3 Examples of Conservation Areas

Conservation Areas have been used by Local Authorities to protect a wide variety of urban spaces. Following are some examples of areas protected by designation as a Conservation Area.

321 For a detailed study on the law relating to trees in the United Kingdom, and on the additional restrictions regarding trees in Conservation Areas, see MYNORS, C. The Law of Trees, Forests and Hedgerows, Sweet & Maxwell, London, 2002, especially pages 577–590.

3.3.1 Stroud Industrial Heritage Conservation Area

Stroud Industrial Heritage Conservation Area was first designated in 1987 and has been extended several times. The Conservation Area came under review in 2006. The Stroud Industrial Conservation Area is a complex construction of various elements in the town, all of which reflect the areas industrial heritage.

According to a draft Conservation Area Statement, presented to the council in November 2008,

> the Stroud Valleys have an important legacy as one of the country's earliest cloth-making areas, a legacy which can be traced back to at least the 14th century. Many mills have since diversified to other industries, and textile manufacture has virtually disappeared in the locality. However, a rich built heritage of former mills, associated buildings and the transport infrastructure (canals, roads and railways) remains. The relationship between built and natural features along the valley bottoms creates a unique and often surprisingly picturesque character, of great variety and contrast.

The Industrial Heritage Conservation Area was originally designated to acknowledge and protect the influences that industry has had in forming the built environment. However, the Industrial Heritage Conservation Area is not exclusively made up of obvious industry. Historically, domestic buildings and small residential enclaves

> have always been closely intermingled with the industrial environment throughout the Stroud Valleys. It should be remembered that many buildings that lack an obviously 'industrial' appearance may actually have links to the industrial heritage of the area: mill workers' cottages, for example, or buildings and structures that relate to the evolving transport infrastructure which was so crucial to the Stroud valleys' industrial progress.[322]

3.3.2 Blackness Conservation Area Dundee

Designated as a conservation area in 1987, according to the local authority:

> Blackness was once regarded as one of the most unsightly parts of Dundee with its mills and workers' tenements packed so closely together. Few areas in Dundee can equal the industrial and social significance of Blackness however, and despite drastic clearance programmes, many fine mills still survive. These, together with the narrow cobbled streets between them, are an accurate reminder of an age now gone. The unique industrial character of this area makes it as worthy of preservation as any of the city's Conservation Areas.
>
> The factories of Blackness developed along the 'Scouring Burn', a natural source of water to power the mills. The burn cannot now be seen, but it still runs underground on a line from Polepark to Guthrie Street.

322 Stroud District Council – Draft of Conservation Area Statement, Volume 1 Summary and Character Overview, Chapter 3 p.3.

> Blackness developed as a prime industrial area in the 19th century when Dundee monopolised the world's jute industry. This area, between Lochee Road and Hawkhill, had the greatest single concentration of mills and industrial buildings in Dundee. The area was one of the cornerstones upon which the prosperity of the city was built. In the 20th century foreign competition put the textile industry into irreversible decline. Mills and tenements became vacant and underused and many were demolished. In the last decade or so, however, Blackness has revived as an industrial area and it had also been the site for the city's first Public Arts Scheme. More recently there has been pressure for residential conversion of many of the large industrial buildings within the area, alongside contemporary housing development.[323]

3.3.3 Fishponds Conservation Area

The Fishponds is an area in the borough of Kingston upon Thames that encompasses a former 18th Century estate. There is a surviving core of the original house, as well as some of the grounds which are now laid out as a 20th century Park. Also surviving are the brick fields, originally developed with terraced cottages in the late 19th century. There is also a group of late 19th century villas close to the park along the old turnpike route.

The council considered that there were certain points of interest that warranted the creation of a conservation area in Fishponds:

- there is a strong slope of land which gives important views to surrounding landmarks;
- the area is of strong historic interest due to its development over the last 250 years, and the survival of important historical features, such as the landscaping features associated with the extraction of brick earth for brick manufacture;
- the surviving part of the original stately home, and the parkland surrounding, is of historical and scenic value;
- Fishponds Park itself has a large collection of mature trees, and an interesting landscape with ponds, streams and steep banks;
- close to the park there are five important Victorian detached Villas;
- there is also a cohesive, late 19th century residential development built on the former brick fields, and a large collection of buildings of architectural interest.

---

323 Blackness Conservation Area overview, produced by Planning and Transportation Department, Dundee City Council, 2005.

Chapter 5

The Protection of Historic Properties in Spanish Law

Protection of historic properties in Spain is a complex mix of state and regional law, both in the area of urban planning and specific laws that deal only with heritage. The Spanish constitution of 1978[324] lays out a complex system of distribution of powers between the state and regional governments which need not concern us here, except to point out that in the area of Heritage Law, the state government has had its right to pass a basic heritage law reaffirmed by the constitutional court through STC 17/1991, of the 31st of January, with the stipulation that it is the regional government, and not the state, that decides which items are to be considered as heritage and which are not. Regional governments can also pass their own heritage laws that, within the framework provided by the state law, develop and improve the level of protection offered.

1 Evolution of Historic Property Protection in Spanish Law

Ashworth and Howery argue that the period from antiquity through to the end of the 17th century was characterized by an irreverent attitude to heritage, with the dismantling of many important buildings, although some measures are taken to protect certain monuments.[325] Benitez de Lugo sees some evidence of an awareness of the need to protect heritage in Vespasian's decree prohibiting the plunder of

324 BOE n. 311 de 29/12/1978.

325 ASHWORTH & HOWERY European Heritage Planning and Management. Intellect Books 1999 p. 35.

sanctuaries, and in the "Partidas" of 1263, where it was established that houses or buildings could not be built close to city walls or churches.[326]

Although towards the end of the 17th century private individuals began to accumulate collections of art, these were chosen more for their unusual nature rather than their cultural significance. Again Ashworth and Howery point out that the 17th and 18th centuries witness the appearance of an appreciation of the value of city and urban spaces, "at least in the eyes of the state, which recognised that buildings and works of art lent prestige to a city, and by extension, to the city authorities, and that they also attract travellers. This is where indirect and subconscious economic appreciation of heritage begins!"[327]

It is however during the Enlightenment that the idea that all historical periods come to an end begins to take form, and by extension, the idea grows that the objects and buildings that moribund civilizations leave behind are witness to the past.[328] During the early 19th century there is a spread and reaffirmation of a growing idea that heritage is a witness to the past, and in Europe the first organisations are created to protect heritage. The *Commisions des Monuments* was created by the French assembly in 1790, and soon after in Spain the *Real Cédula de S.M. y Señores del Consejo por la qual se aprueba y manda observar la instrucción formado por la Real Academia de recoger y conservar los monumentos antiguos descubiertos o que se descubran en el Reyno*,[329] while in 1806 Carlos V passed a law prohibiting the export of antiquities from Spanish territory.

Alegre Avila considers that the first real legal regulation of heritage in Spain appears under the Borbon dynasty where, under the patronage of the Academy of Fine Arts of San Fernando, a series of laws are passed protecting specific types of antiquities, monuments and works of art.[330] The Real Orden of the 24th of July 1844, which approves the rules that should be followed by the Provincial Commissions of

---

326 BENITEZ DE LUGO y GUILLEN, El Patrimonio Cultural Español, Comares, Granada, 1995, p. 14.

327 ASHWORTH & HOWERY European Heritage Planning and Management op. cit. p. 37.

328 ASHWORTH & HOWERY European Heritage Planning and Management op. cit. p. 38.

329 Real Cédula of the King and the Gentlemen of the Council by which they approve and order the observation of the instruction formed by the Real Academy to collect and conserve the ancient monuments discovered or that may be discovered in the Kingdom.

330 ALEGRE ÁVILA, Evolución y régimen jurídico del Patrimonio Histórico, Ministerio de Cultura, Madrid, 1994, p.41.

Historic and Artistic Monuments "is, without doubt, the first major regulation of cultural heritage".[331]

Gonzalez-Ubeda Rico points out that these Commissions, which are placed under the control of the Central Commission of Monuments in the Real Decreto of the 15th of November 1854, are the "first hint of the creation of an administrative organisation dedicated to protecting cultural heritage".[332]

The Central Commission of Monuments was abolished in 1857 by the Moyano Law, which, for Garcia de Enterria, integrates in "a unified vision, educational and fine arts policies, and places the control of cultural heritage in the hands of the Real Academia de San Fernando, as the administrative agency hierarchically superior to the Provincial Commissions of Historic and Artistic Monuments, which continue to be the main reference point for those laws passed in the 19th century".[333]

Alegre Avila, for his part, feels that the 19th century lacks a true legal code to protect historic and artistic works, being made up by a "body of dispositions, made up of organic or competential law, whose fatal weakness is that they were passed only to deal with specific, concrete problems that appeared from time to time".[334]

Amongst the most relevant laws passed at the beginning of the 20th century are the Law of Excavations and Antiquities of 1911 which aims to protect "all works of art and industrial products dating from prehistoric and ancient times and the middle ages", and the Law of Monuments of 1915 which defines historic monuments as "those that have historic and artistic merit, whatever their style, which, whether considered as a whole, or partially, are deemed to possess such merits in an administrative process, which can be requested by any individual or corporation, and which must be included in a catalogue created by the Ministry of Fine Arts".[335] This Law therefore includes two important elements – the creation of an administrative process to declare the existence of specific cultural properties, and the posterior registration of the item of heritage in a specially created catalogue, both of which in one form or another continue to the present day.

331 ALEGRE ÁVILA, Evolución y régimen jurídico del Patrimonio Histórico op. cit. p. 46.

332 GONZÁLES-ÚBEDA RICO, Aspectos jurídicos de la protección del Patrimonio Histórico-Artístico y Cultural, Ministerio de Cultura, Madrid, 1981, p. 93.

333 GARCÍA DE ENTERRIA, "Consideraciones sobre una nueva legislación del Patrimonio artístico, histórico y cultural", REDA, number. 39, 1983, p. 578.

334 ALEGRE ÁVILA, Evolución y régimen jurídico del Patrimonio Histórico. op. cit. p. 47.

335 Article 1 of the Law of Monuments of 1915.

The Decree of the 9th of August 1926 represents an important development in the legal techniques used to regulate heritage, establishing a strong connection between the regulation of historic buildings and urban planning. In this law, the administrative process could be initiated both by private individuals and those organisations enumerated in the law.

The declaration of an object as a part of the nation's "artistic treasure" has important consequences in this law. According to its article 7, the conservation, protection and custody of architectonic monuments is considered to be of "public interest". In the case of buildings, the declaration implies that not only the building itself, but also all those parts of the surrounding area that compliment it, become part of the "nation's land". As such, they cannot be demolished, either fully or partially, without the express consent of the Ministry. Also, crucially, the fact that the heritage forms part of the artistic treasure of the Nation means that its owner, possessor or user has the obligation to conserve the object for the Nation. Article 7 of the Law of the 9th of August is considered as a direct predecessor of article 36 of the current Spanish law, which allows for the subsidiary action of the state, either by expropriation or by "intervention in works of transformation or reform proposed by the owner, so that the typical aspect and characteristics of the monument are not altered".

One of the most outstanding aspects of the 1926 law is described by Fernandez Rodriguez as the creation of a bridge between "art and urban planning".[336] Article 26 of the 1926 law obliged local councils to create topographic plans which showed which buildings

> are subject to restrictions on their development, marking clearly those artistic and historic buildings, places, streets, squares and areas in which no development can take place without the authorisation of central or provincial authorities. Building works will take into account these limits and plans. The local council cannot carry out any public works or use the faculties of expropriation without authorisation of those institutions involved in the declaration of a city or town as 'artistic treasure' or by the Ministry of Fine Arts.

Evidently, these topographic plans serve as a precedent for the special protection plans for Historic centres included in article 20 of the actual Spanish law.

---

336 FERNANDEZ RODRIGUEZ "Legislación española sobre protección del patrimonio artístico. Balance de cara a su reforma", Revista de Derecho Urbanístico **60**, 1978 op. cit. p. 14.

The next law to deal with heritage was the Law of Defence, Conservation and Promotion of the National Artistic and Historic Heritage of 1933, regulated by Decree of the 16th of April 1936 and modified by Decree 1545 of 1972 which was in force until the passage of the actual Spanish Heritage Law in 1985. According to this law, National Heritage was comprised of all those movable and immovable objects[337] which include the values described in the law and which are at least 100 years old. Younger buildings could only be declared as heritage where their artistic or historic value was indisputable. However the 100 year rule was taken as a norm in the application of the law.

The 1933 law uses the technique of a formal declaration of an object as a Historic Artistic Monument. This formal declaration is the final result of an administrative procedure which includes some elements which are common to all administrative procedures, as well as some which are specific to this law (commencement of the procedure, documentation, guarantee of a public hearing of interested parties, necessary reports, notification of the declaration, inclusion in a national catalogue). The law places considerable emphasis on the heritage catalogue, and dedicates a whole section to the creation of the Inventory of Historic-Artistic Heritage. The inventory also unified the Catalogos Monumentales and the Ficheros Del Arte Antiguo.[338]

Under the 1933 law, protection of the historic building is in force as soon as the administrative procedure begins, with a prohibition on demolition or reform of the monument (including work already started). Where imminent danger of collapse existed, the local conservation architect would have to become involved. Once the monument was formally declared as heritage, it was subject to a series of measures to preserve and defend those values which led to the monument being declared as such. Alegre Avila points out that "this collection of limits and positive obligations, counterbalanced by a series of benefits, mean that the building, when formally declared, is included in a special legal category or statute – National Monuments – inspired by the basic idea of the untouchable nature of the monument".[339] Muñoz Machado points out the fundamental nature of

337 The term immovable object (bien inmeuble) is not analagous in spanish to 'building', but also includes and other thing considered as fixed – for example gardens, statues fixed to the ground, fountains, areas of land, bridges, etc.

338 MARTINEZ GARCIA, "Inventario y catalogación del Patrimonio Histórico. Andalucia como marco" in Patrimonio Cultural y Derecho 7 2003 p. 59.

339 ALEGRE ÁVILA, Evolución y régimen jurídico del Patrimonio Histórico op. cit. p. 164.

article 19, which will "serve as a base for all the anti-reconstruction measures, and which is the predecessor, although much more flexible, of article 39.2 of the present law".[340]

The administration could also intervene directly when the owner did not fulfil a duty of care to the monument. The law also subjected most works on historic buildings to administrative authorization. Article 23 of the 1933 law lays out that "the owners, those in possession and those using historic and artistic monuments cannot carry out any type of work whatsoever in the monument without the perceptive authorisation from the Junta Superior del Tesoro Artístico, which will require the consent of the area architect".

Other important elements in the 1933 law are the establishment of the right to visit the monument (article 29, predecessor of article 13.2 Spanish Heritage Law), the establishment of a framework for the transmission from person to person of the monument, which gives the state a preferential right in the sale of the monument, and finally the law also takes into account the possibility of expropriation of buildings which impede the contemplation of a monument.

2 The Protection of Historic Properties in the Spanish Constitution

Of the three countries object of study in this publication, only two, the United States and Spain, have a written constitution, and of these, while the United States Constitution touches tangentially on some issues of interest, such as the idea of unlawful taking of property, Spain is the only one that refers directly to a right of access and of protection of cultural heritage. Spain is a constitutional monarchy, and as such, all laws have to be interpreted according to the 1978 Constitution.

It is, therefore, of interest to scholars in different legal systems to understand the philosophy behind the "rights" given to Spanish citizens in their constitution, and the obligation that the text places upon public powers in that country.

2.1 The Right to Heritage in Article 46 of the Spanish Constitution

Included within those essential values that make up modern society is the conservation and enjoyment of those concrete things and those

340 S. MUÑOZ MACHADO, La resurrección de las ruinas, Civitas, Madrid 2002, p. 69.

cultural manifestations that compose the cultural identity of a people, and the guarantee of global access to them to all citizens. These basic values are included throughout the Spanish constitution, and in the case of heritage, in article 46, strongly connected to article 44.1 CE[341] (the right to culture) and article 45 CE (the right to an adequate environment),[342] as well as the right to education and the case law relative to the social right to culture developed by the Supreme Court.[343]

Perez Luño defends the structural order of the Spanish constitution, with heritage being placed after the article announcing a general right to an adequate environment rather than after the article declaring a right to equal access to culture. For this author, "while there is an undoubted connection between access to culture and the protection of heritage, a right to have our heritage protected can be understood as a subpart of environmental protection. It is, therefore, correct to place the constitutional recognition of heritage after the recognition of a right to an adequate environment".[344] For this author, there are various constitutional precepts whose aim is to protect people from the "economic leviathan", or, in other words, blind domination of the natural and human world by brutish economic laws that produce personal alienation. For this author, article 46, which protects the conservation and enrichment of cultural heritage, and which is part of a wider concept of environmental protection, is situated in a series of constitutional precepts that "tend to discipline the economic use of commodities in order to avoid antisocial use and to guarantee the satisfaction of our basic necessities".[345]

---

341 Article 44.1 CE establishes that "Public authorities will promote and protect access to culture, to which all citizens have a right".

342 Article 45 establishes that

- everyone has the right to enjoy an environment suitable for the development of the person, as well as the duty to preserve it;
- the public authorities shall watch over a rational use of all natural resources with a view to protecting and improving the quality of life and preserving and restoring the environment, by relying on an indispensable collective solidarity;
- for those who break the provisions contained in the foregoing paragraph, criminal or, where applicable, administrative sanctions shall be imposed, under the terms established by the law, and they shall be obliged to repair the damage caused.

343 This jurisprudence is developed, amongst others, in the following cases: STS of the 9th of April 1984, STS of the 13th of April 1984, STS of the 11th of February 1985, STS of the 17th of May 1985, STS of the 3rd of October 1985, STS of the 3rd of October 1986, STS of the 8th of May 1987, STS of the 19th of November 1991 and STS of the 6th of April 1992.

344 PEREZ LUÑO, Derechos humanos, estado de Derecho y Constitución, Tecnos, Madrid, 5ª ed, 1995, p. 283.

345 PEREZ LUÑO, op. cit. p. 465.

Martín Mateo, on the other hand, argues against placing heritage protection together with the right to an adequate environment. For this author, the legal techniques that apply to the environment and heritage are so different as to make any attempt at assimilation confusing. However, this same author does agree that there are notable similarities in the two concepts that justify the wide powers of intervention given to the administration.[346]

Article 46 CE, whose origins are to be found in the Real Cedula of the 6th of June 1803 and article 45 of the Republican Constitution of 1931, literally states: "The public authorities shall guarantee the preservation and promote the enrichment of the historical, cultural and artistic heritage of the peoples of Spain and of the property of which it consists, regardless of their legal status and their ownership. The criminal law shall punish any offences against this heritage".

There are three main elements in this formulation. In the first place, a global concept that avoids entering into detail on what cultural values may be important at any given time to society. Barrero Rodriguez writes that "an attempt is made to bring into the legal world concepts such as culture, art and history, all of which are notions subject to those values and judgements that society bestows on the work or the object at any given historic moment".[347] The text of article 46 is therefore intentionally vague, aiming to establish and underline "an extended and unifying concept in which a diversity of ideas and categories of heritage may find a home".[348]

The second major element of article 46 is its dynamic nature. Rather than simply declaring the protection of heritage, the article assigns a positive role to public authorities – that of promoting and enriching cultural heritage. While not specifying the exact nature of this role, it does represent a radical departure from the pre-constitutional situation, where, according to Fernando Rodriguez, the role of public authorities was decidedly non-interventionist, with concrete and specific limitations on the state's role imposed by an idea

---

346 MARTÍN MATEO, R. Prologue to ALONSO IBÁÑEZ, El patrimonio histórico. Destino público y valor cultural, Civitas, Madrid, 1992, p.s 23–24.

347 BARRERO RODRÍGUEZ, La ordenación jurídica del Patrimonio Histórico, Madrid, Civitas, 1990. With respect to the possible redundant nature of this article, given that other articles in the constitution deal with similar concepts, PEREZ LUÑO, believes that, while it is intimately related to other ideas, it does possess a meaning, in some cases wider and in some cases more restrictive than those articles that seem reiterative, which in this authors opinion, justifies its separate annunciation. PEREZ LUÑO, Derechos humanos, estado de Derecho y Constitución, op. cít. p. 480.

348 RUIZ-RICO, G. "*La disciplina constitucional del Patrimonio Histórico en España*", Patrimonio cultural y Derecho **4**, 2000, p. 29.

of culture still dominated by private ownership.[349] As Alonso Ibañez points out "the actual Spanish constitution requires a much more ambitious policy: to be promoters and agents of culture".[350]

Perez Luño also emphasizes the dynamic role expected of public authorities in protecting heritage, which is due to the

> conscience acquired in recent years that the mere conservation of heritage is not sufficient. It affirms a much more active policy than the simple defence and restoration of cultural and artistic works. The active role that corresponds to the public authorities in the protection of cultural heritage as a part of culture implies not only the promotion of artistic works, but also requires a real and effective policy that allows heritage to be accessible to society as a whole.[351]

The third major element of the concept of heritage included in article 46 is what has been called the "democratization of culture", creating a connection between the social function of cultural objects and other principles and rights, such as the freedom to full personal development, culture and environment, which, as Orozco Pardo points out "signifies certain spiritual values, before which the full right to private property is limited, and is therefore subordinated to the public interest".[352] Developing this line of interpretation, the STS of the 13th of April 1981 explains that

> the social right to culture which now transcends national law and is object of an increasing international regulation, and which is now understood, on a spiritual level, to be part of a collective heritage of humanity, has certain essential qualities and is of such importance, that legislation to protect it is absolutely essential in a way that is most favourable to the finality of the full conservation of cultural heritage.

Cano Murcia, for his part, feels that the understanding of cultural heritage in function of its collective spiritual nature is an expression of diffuse or public interest that can be brought into play through the class actions provided for in article 125 CE.[353]

As to the full legal nature and reach of the right to cultural heritage included in article 46, Garcia Macho explains that, although all the

---

349 FERNÁNDEZ RODRÍGUEZ, T.R. "Legislación española sobre protección del patrimonio artístico. Balance de cara a su reforma", op. cit, p. 13.

350 ALONSO IBÁÑEZ, M.R. El patrimonio histórico. Destino Público y valor cultural, op.cít., p.s161 ss.

351 PEREZ LUÑO, A. Derechos humanos, estado de Derecho y Constitución, Tecnos, Madrid, 5ª ed, 1995, p. 491.

352 OROZCO PARDO, G. & PEREZ ALONSO, E. La tutela civil y penal del Patrimonio Histórico, cultural y artístico, McGraw Hill, Madrid, 1996, p. 42.

353 CANO MURCIA, M. "El interdicto de obra ruinosa y el procedimiento de declaración de ruina", La Ley, **2714**, mayo, 1991

rights affirmed in articles 39–52 CE should be developed by the lawmaker uniformly, that is not actually the case. According to this author, the effective concretion of each one of these principles will depend on the precise drafting of the constitutional precept. In those cases where the Constitution uses the term "all people" or "all Spanish people have a right to", the lawmaker has a more intense obligation to develop the right into firm law, whereas, in those articles where the text follows the formula of "public powers will promote/maintain/ensure", there is a weaker obligation. In this sense, the constitutional mandate of rights to health, environment, culture, housing and heritage are not fundamental rights in the strictest sense, as demonstrated by the drafting of article 53.3 CE.[354]

Article 53.3 CE declares that "Recognition, respect and protection of the principles recognized in Chapter 3 shall guide legislation, judicial practice and actions by the public authorities. They may only be invoked before the ordinary courts in accordance with the legal provisions implementing them". This gives an enormously important role to public administration, which, as a public power, has an essential role in the development of these guiding principles. It is, however, the legislative branch that must take the first step, as article 53.3 establishes the necessity that a law develops these principles by stating that "they can only be invoked before the ordinary courts in accordance with the legal provisions implementing them."

2.2 The Concept of "Social Function" as a Limiting Factor on Individual Property Rights

As previously mentioned, article 46 guarantees the protection of items of cultural heritage "whatever their legal and property status". As Abad Liceras points out, the public or private ownership of heritage is completely and utterly irrelevant as a fundamental limiting factor in its legal protection. The principle element to be considered here is the convergence of two types of interest: one of a general, public nature (represented and defended by the administration), and the other private. When both interests come into conflict, according to Spanish constitutional theory, the public interest must prevail.[355]

354 GARCIA MACHO, R. "De los derechos sociales a los principios rectores", Separata del libro La Declaración Universal de los Derechos Humanos en su 50 aniversario, Editorial Bosch, Barcelona, 1998, p. 365.

355 ABAD LICERAS, J.M. "El principio jurisprudencial del derecho social a la cultura: análisis y manifestaciones" in Patrimonio Cultural y Derecho 7, 2002 p. 33.

The possibility of limiting property use using the legal method of the social function of property is declared in article 33 CE, and as part of our study on the regulation of cultural heritage, the content of property rights and the conflict between private use and the social function of property have to be understood.

According to article 33:

1. the right to private property and inheritance is recognized;
2. the social function of these rights shall determine the limits of their content in accordance with the law;
3. no one may be deprived of his or her property and rights, except on justified grounds of public utility or social interest and with a proper compensation in accordance with the law.

Perez Luño feels that the most significant aspect of article 33.2 CE is its express use of the term "social function" as a limit on the right to property and inheritance.

> Nevertheless, the whole notion of the 'social function' is not completely unambiguous. The idea is an important element within the principles of a social and democratic state, underpinning those very values that inform the statute of fundamental rights and the very concept of an economic Constitution. The concept of 'social function' is a factor that delimits, and is delimited by these criteria, without being weakened as a relevant constitutional mandate towards the legislative branch.[356]

In this very line of argument, STC 37/1987 of the Spanish constitutional court establishes that

> fixing the 'essential content' of the concept of private property cannot be done through an exclusive consideration of the subjective or individual rights that underline it, but must include a reference to property's social function, understood not simply as a limiting factor in the definition or exercise of property rights, but rather as an integral part of the whole concept of private property. Individual use and social function together define the true meaning of ownership.

For its part STC 227/1988 established that the ownership rights of heritage owned by private individuals sanctioned by article 33 of the Constitution

> is not absolute, given that article 128.1 CE establishes that 'The entire wealth of the country in its different forms, irrespective of ownership, shall be subordinated to the general interest', and article 45.2 obliges public authorities to ensure 'rational use of all natural resources with a view to protecting and

356 PEREZ LUÑO, A. Derechos humanos, estado de Derecho y Constitución, Tecnos, Madrid, 5ª ed, 1995, p. 441.

> improving the quality of life and preserving and restoring the environment, by relying on an indispensable collective solidarity'. A systematic interpretation of these constitutional precepts does not allow us to accept the idea that any legal regulation of natural resources must place first criteria that avoid sacrificing individual rights or interests...The connection between ownership, social function and public interest is based on society's values, and these are not limited to the simple notion of economic development, but include other values, including fundamental values of humanity and cultural development. All this because the rights and obligations of a citizen are included in a system based on social, political, economic and cultural rights.

For Alvarez Alvarez, it is evident that it is in the very nature of those things that make up heritage that the social function inherent to them means that they should be handed back, in some way, to the society that created them, without negating totally the fact that heritage may be privately owned.[357] According to Orozco Pardo and Perez Alonso

> it can be clearly inferred that the constitutional idea of 'social function' as a conditioning factor to ownership of those items that integrate Spanish historic and cultural heritage only makes sense through the promotion of public access to them. The right of ownership of heritage has to be exercised in such a way as to promote cultural enjoyment by society. Heritage that is protected by law is made up of those socially and culturally relevant items that make up the historic, cultural and artistic heritage of the Spanish people.[358]

In conclusion, the so called "social function" of ownership can be understood in terms of making heritage accessible to citizens, in such a way that the right to private ownership has to be understood as limited by the need to permit collective access and enjoyment.

3 Spanish Heritage Law

The constitutional issues considered above have been regulated through Law 16/1985 of the 25th of June.[359] Ruiz Rico feels that "it is a law that solidifies the principles and criteria of the constitution, although it not always using the same criteria. The Spanish legislator has distanced itself from the terminology used in the Constitution in laying out the concept and object of the law: 'Spanish historic heritage', rather than 'Spanish historic, cultural and artistic heritage'". According to this author "the lawmaker's intention has been not only

357 ALVAREZ ALVAREZ, J.L. Estudios sobre el Patrimonio Histórico Español, Civitas, Madrid, 1989, p. 68.

358 OROZCO PARDO, G. & PEREZ ALONSO, E. La tutela civil y penal del Patrimonio Histórico, cultural y artístico, op. cít., p. 146.

359 BOE n. 155 de 29/6/1985.

to establish in one law general principles relating to the protection of historic heritage, but also to elaborate a legal system for each different type of heritage (archaeological, anthropological, documental and bibliographic) within the context of a single, unitary conception of culturally relevant objects".[360]

The law establishes the following categories of heritage:

- a general category: objects included in Spain's historic heritage, made up of all those items that have historic, artistic, scientific or technical value and that "are part of Spain's contribution to universal culture";
- a specific category, that of culturally relevant objects (Bienes de Interés Cultural), that is a subset of the first category, but which enjoys a greater protection. Bienes de Interés Cultural are characterized by having unique cultural characteristics that differentiate them from heritage in general. As well as the special regime of guarantees for these objects, a whole range of fiscal measures are established, and access rights for all citizens are guaranteed;
- Individual categories that have their own specific regulation: archaeological, anthropological, documentary and bibliographic heritage.

Article 1.2 describes in general terms those items that make up Spain's historic heritage in the following terms: "The Spanish Historical Heritage is made up of movable and immovable objects of artistic, historical, paleontological, archaeological, ethnographic, scientific or technical interest. It also comprises documentary and bibliographical heritage, archaeological sites and areas as well as natural sites, gardens and parks having artistic, historical or anthropological value". Article 1.3 completes this description by specifying as a subset "the most relevant property forming part of the Spanish Historical Heritage, which shall be inventoried or declared of cultural interest as a 'Bien de Interés Cultural'".

Spanish Heritage Law thus defines what objects are worthy to be protected, promoted and passed on to future generations, in virtue of their "interest", albeit historic, artistic or as defined in any of the categories in the law. It is a concept that is vague, ambiguous and indeterminate. The problem the lawmaker has to resolve is common to

360 RUÍZ-RICO, G. "*La disciplina constitucional del Patrimonio Histórico en España*", Revista de Patrimonio Cultural y Derecho **4**, 2000, p. 59.

all heritage laws – a definition that is too tight will be too binding, and will ossify a definition of culture, which, as we have seen, is a concept characterized by its very fluidity. For Giannini, "culture can be nothing if it is not an open idea, a notion to which legal regulation cannot give content based on concise legal definition, but that must be defined by reference to other non-legal disciplines, which means therefore that the legal concept is defined by non-legal experts".[361]

The concrete definition of what is to be protected by heritage law has great practical significance, given that the declaration of a Bien de Interés Cultural brings down upon the owner a stringent set of obligations and restrictions. The declaration does not create a personalised regime, nor does it create the cultural value in itself, but rather it just confirms that that historical value does indeed exist. That determination comes from an act of evaluation, which is, of course, loaded with personal judgements. The declaration of a Bien de Interés Cultural is the final act in a long and complex administrative process in which the public authority has no choice but to work with an ambiguous concept.

This act of evaluation means that the administration must take into account the specialized criteria that come from disciplines considered very distant from the law, although the existence of value judgements in the administrative act does not, of course, prevent judicial control of the administrative decision.

Some authors warn of the arbitrary nature of the process. Perez Luño feels it is essential that the discretionary powers enjoyed by the administration do not degenerate into arbitrary decision making through the evaluation of those ambiguous concepts of historic and artistic value that are intertwined with our concept of heritage. He does point out, however, that the use of a more restrictive definition can be even more dangerous: "history shows us that many of the most important features of our heritage have been systematically destroyed. In a scenario where we want to preserve not only our most important monuments, but also their surrounding areas, the use of restrictive criteria could be used as an excuse for the continued destruction of elements with cultural relevance to our society". In consequence, "taking into account the constitutional framework which forces us to consider culture in the context of a changing society (art 3.1 CE) and the limiting factor of the social function of property (article 33.2 CE)

361 GIANNINI, "*Beni culturali*", Rivista trimestrale de diritto publico, 1976 **1**, p. 8.

we can only hope for firm action from our courts to protect our cultural heritage".[362]

3.1 The "Bien De Interés Cultural"

In the case of immovable objects (for example buildings, areas of a town or city and landscapes) the most relevant form of protection is the Bien de Interés Cultural, a category that includes only the most relevant elements of heritage, each of which must be declared as such by a formal administrative act. Those Bienes de Interés Cultural that are immovable can be declared as: Monuments, Historic Gardens, Historic Sites, Historic Areas or Archaeological Areas.

The Bien de Interés Cultural, as pointed out by Lopez Bravo, is the central element of Spanish Heritage Law, and has its roots in the Italian concept of bene culturale, elaborated by Italian scholarly legal opinion in the second half of the 20th century[363]. The introduction to the law recognises the fundamental role that the Bien de Interés Cultural plays in heritage legislation, pointing out that "within Spanish Historical Heritage, and for the purpose of granting greater protection and safeguarding, the category of "Bien de Interés Cultural" takes on special value, covering that movable and immovable property which has the greatest need of such protection. This category implies special measures which the Law lays out in accordance with the type of property covered."

3.1.1 The Formal Declaration of Historic Buildings as "Bien De Interés Cultural"

Responsibility for declaration of a property as a Bien de Interés Cultural corresponds to the regional government,[364] and for the

---

362 PEREZ LUÑO, A. Derechos humanos, estado de Derecho y Constitución p. 487.

363 Vid. LOPEZ BRAVO, "*Los bienes culturales en el Derecho estatal y autonómico de España*"in *Patrimonio Cultural y Derecho*, number 3, 1999, p. 11. The term bene culturale enters into Italian Law following the deliberations of the Franeschini Commision, which defined it as "that object which is a material witness of culture, because it embodies values of our civilization".

364 Artículo 11.1 REAL DECRETO 111/1986, de 10 de enero, de desarrollo parcial de la Ley 16/1985, de 25 de junio, del Patrimonio Histórico Español (BOE nº 24 de 28 de enero de 1986), which establishes "Artículo 11. 1. Corresponde a cada Comunidad Autónoma incoar, de oficio o a instancia de cualquier persona, los expedientes para declarar de interés cultural los bienes de titularidad pública o privada que se encuentren en su ámbito territorial, sin perjuicio de lo dispuesto en el apartado siguiente.2. Corresponde al Ministerio de Cultura incoar, de oficio o a instancia de cualquier persona, los expedientes para declarar de interés cultural los bienes integrantes del Patrimonio Histórico Español adscritos a servicios públicos gestionados por la Administración del Estado o que formen parte del Patrimonio Nacional".

purposes of clarity, we will here lay out the procedure that has been adopted in our own regional government – the Comunidad Valenciana (Valencia) through law 4/1998,[365] modified substantially by law 5/2007.[366]

In Valencia, the declaration of a Bien de Interés Cultural is carried out by a Valencian Government Decree, proposed by the Department of Culture, Education and Science. The administrative procedure can be started by the administration of its own accord, or as a response to a request by an individual citizen. The beginning of the administrative procedure must be notified to all interested parties and to the local council. Where an administrative procedure to protect a particular property has been requested by a private individual, the administration may decide not to go forward, notifying the person who has requested the declaration within three months, with a full explanation of the reasons used to take the decision.

The administration cannot initiate the procedure without favourable reports from at least three of the following Valencian institutions: The Valencian Cultural Council, the Royal Academy of Fine Arts of San Carlos, the Universities of the Valencian Community and the advisory councils for collections and libraries of Archaeology and Palaeontology. The report has to be emitted within three months, after which, if it has not yet been issued, it is understood by default to be favourable. If one of the reports is negative, there must be at least two reports that expressly confirm the administrative procedure (i.e. that are not favourable due to default). The opening of the process must be published in the official daily publication of the Valencian government (Diario Oficial de la Generalitat Valenciana – DOGV). The official Land Registry must also be notified of the procedure.

The local council must be given a hearing, and a period of public information of one month is opened. Once the administrative procedure is begun, as a preventive measure, the heritage under consideration enjoys all the protection given to historic properties that have already been declared as such, and all municipal licences and zoning plans that may affect the property are immediately suspended.[367] This suspension can be lifted by the department of

365 LEY 4/1998, de 11 de junio, de la Generalitat Valenciana, del Patrimonio Cultural Valenciano.

366 LEY 5/2007, de 9 de febrero, de modificación de la Ley 4/1998, de 11 de junio, del Patrimonio Cultural Valenciano.

367 New redaction of Article 33 given by Law 5/2007.

culture if it is felt they do not affect the essential nature of the property in question.

Once the administrative procedure is initiated, it must be terminated within fifteen months for immovable objects, one year for movable objects and two years for immaterial heritage, which is calculated not from the date the administrative procedure is requested, but rather from the date of initiation of the administrative procedure. This time period can be extended by twenty months where the heritage in question is especially complex. The administrative procedure is considered null and void if a declaration is not made in the thirty days following the end of the limits specified above, and once an administrative procedure is voided, it cannot be restarted within the next three years. This restriction does not apply where the petitioner is the owner of the heritage or one of the advisory bodies involved in the administrative procedure.

The Decree that declares the Bien de Interés Cultural must lay out clearly the specific qualities of the item of heritage, containing a detailed description of all aspects of the property. The declaration must also make clear in which sub-category it is declared. The declaration must also determine the geographical limits of the item of heritage and its area of influence, as well as its relationship with the surrounding area. It will also list those integral parts, belongings and accessories to the object that are considered to be included as part of the declaration. If the declaration is of a Historic District, the declaration must clearly specify which buildings are included, describing all of them in detail.

The declaration must be inscribed in Section 1 of the Valencian Heritage Register and the State Heritage Register. All interested parties must be notified, including the local council. The declaration must also be published in the Boletín Oficial del Estado, and the Diario Oficial de la Generalitat Valenciana. All immovable property included in the decree must be inscribed as such in the property register.

3.1.2 The Different Categories of Bien De Interés Cultural

A) Monuments

Article 15.1 of the law provides that “Immovable property comprising architectural or engineering work or works of colossal sculpture shall be Monuments provided they are of historical, artistic, scientific or social interest”. The expression “architectural” refers to any type of building, whereas the expression “engineering” refers to objects such as bridges. The expression “colossal sculpture” is more difficult to tie down, although it is generally understood to mean those human works that can be integrated into the common artistic concept of "sculpture" and which are fixed to the ground.

Examples of Bienes de Interés Cultural that have been declared using the category of Monument are:

- The Flour Factory "La Constancia" in Murcia was declared as a Bien de Interés Cultural through Real Decreto 50/1992 of the Regional Government of Murcia.[368] According to the decree, the Constancia constitutes one of the most important examples of industrial architecture existing in the city of Murcia. Its position opposite the Martillo del Palacio Episcopal and the Molinos de Segura and the Plaza de Camachos give it special significance as part of an area that defines the image of a section of Murcia's urban structure. It also represents an important step in the modernization of the sector of flour making, and following the abandoning of the hydraulic mills on the banks of the River Segura, it is an essential link in our understanding of our industrial heritage. The decree includes in the declaration the building itself, plus all the relevant machinery that is still inside.[369]
- The Royal Artillery Factory in Seville, declared through Real Decree 1266/2001 of the 16th of November. The decree states that the Monument is an important part of Seville's industrial architecture, closely related to Spain's maritime expansion. The factory follows a typical conception of architecture used in the Royal Factories: "with a great sense of unity between the different spaces in which work was carried out, incorporating schemes of urban morphology in its interior with a projection onto the surrounding areas. The factory was built in a late baroque style with elements of neoclassicism, very much in the

368 Boletin Oficial de la Region de Murcia (BORM) 5th of June 1992.

369 The decree details the machinery as follows: "tres tornos hexagonales, una cepilladora de salvado, una cepilladora normal, ventiladores, los filtros de limpieza de aire modelo J.L., cuatro Plansichter, dos sasores, un torno, un molino de remolido de salvado, rociadores de trigo y depósitos de reposo, ochenta y cinco conducciones de subida de cereales de madera de pino rojo de 3,5 metros de largo, cincuenta y cinco conducciones de unos dos metros de largo, de la misma madera, cincuenta cintas transportadoras de cereales de lona con recipiente de hojalata, siete básculas industriales, cuatro máquinas de moler BSGDG, trece rodillos de hierro" correspondientes a las máquinas anteriores, y un motor eléctrico y maderas de distintas clases".

style of other similar large buildings built by military engineers".[370]

B) Historic Gardens

Historic gardens are defined as "a delimited area resulting from organisation by man of natural elements, sometimes complemented by constructions, and considered of interest because of its origin or historical past or its aesthetic, sensory or botanical values". The historic garden is differentiated from a "picturesque area" – an area of vegetation of outstanding beauty, by the presence of human intervention.

C) Historic Districts

The category of Historic District has caused interpretational difficulties in the past. Cantucci explains that we are faced with an abstract notion which cannot be expressed as a "thing". It is rather, an entity that is made up of things which are better described as being "related to each other". The relationship between the buildings in the District is defined by the aesthetic and positional relationship with other buildings. The object of protection is the artistic/aesthetic identity of the area, which, of course, directly affects ownership rights on each element in the area.[371]

The protection of a district of buildings first appears in Spanish Law in the Heritage Act of 1926 which protects "those built districts whose protection and conservation are necessary in order to protect those physical, artistic and picturesque aspects that characterize Spain".[372] This wording, which has been widely praised by recent authors,[373] was thrown out in the 1933 law, which placed particular emphasis on individual buildings, referring only sporadically to "districts" or "areas", with a definition that included "streets, squares, barrios, walls, fortresses and ruins outside of towns that can be included in the categories of squares, streets, barrios or historic-artistic areas because of their particular beauty".

In the present law, article 15.4 defines Historic District as:

---

370 BOE number 287, 30th of November 2001 p. 44236. The monument was declared by the national government as it was under state management at the time.

371 CANTUCCI, La giuridica dell cose d'interesse artístico e storico, Cedam, Padova, 1953 p. 151.

372 Article 3 of the Real-Decreto of 1926.

373 See for example, FERNANDEZ-RODRIGUEZ, "Legislación española sobre protección del patrimonio artístico. Balance de cara a su reforma", op. cit. p. 13 ss.

> a group of immovable properties forming a continuous or dispersed unit of settlement, covered by a physical structure representing the development of a human community in that it testifies to their culture or constitutes a value for public use and enjoyment. A historical district is also any individualised group of properties included in a larger population unit having the same characteristics and that can be clearly delimited.[374]

For Alonso Ibañez this definition is ample and open with the important element being its ability to bear witness to human culture or of having values of use or enjoyment, "that is: that they are historic-urban documents or are integrated naturally, forming a single, representative unit, independent of whether each individual element included in the area has its own special value".[375] The limiting factor is not the nature of the individual buildings that make up the district, rather the value of the district itself as a witness to human culture, independently of the nature of each individual element.

Given that a Historic District only exists as a collection of buildings and other objects, the determination of the level of protection that is to be given to each individual element, and whether that differs from the level of protection given to the district as a whole, is a complex legal issue. If we understand the Historic District as an immaterial thing, defined by Diez Picazo as those realities, which, although they lack corporal existence, and being a creation of the human spirit, are valued by the legal system as an whole, then the logical conclusion is that the protection applicable to the Historic District applies to all those things in the district that contribute to those values, insofar as they are determinate material realities. The level of protection is not given indiscriminately to all properties in the district, only those that actually contribute to whatever makes it worthy of protection.[376]

The protection offered to those buildings declared as Bienes de Interés Cultural extends beyond the buildings themselves to the area that surrounds them, but which do not have the same historic-artistic value as the building itself. These are described in the law as the "entorno" – the surrounding area. This area is not only protected in the Spanish Heritage Law, but also in the various urban planning laws. The Spanish Heritage Law forces the public administration to define the surrounding area of an immovable Bien de Interés Cultural, establishing a regime of administrative intervention in the area, which

374 Article 15.3 Spanish Heritage Law.

375 ALONSO IBAÑEZ, M.R. El patrimonio histórico. Destino público y valor cultural, op. cit., p. 62.

376 DÍEZ PICAZO, Fundamentos de Derecho Civil op. cít., p. 124.

is, as would be expected, less intense than the regime for the Bien de Interés Cultural itself, given that the buildings in the surrounding area do not have necessarily intrinsic cultural value, but are valued only insofar as they compliment the heritage itself. In this sense, STS of the 8th of March 1991, referring to a case brought under the 1933 law points out that

> by definition, an 'historic district' and a 'surrounding area' are not the same thing, given that the latter is protected as a preventive measure, only insofar as it protects the values of the Historic District. In other words, the end in itself is the protection of the Historic District, and the protection of the surrounding area is only a means to this end. Therefore, the end in itself, and the means to the end, require different administrative regimes and actions.
>
> While the protection of the buildings in the Historic District is extreme, in those buildings that are in the surrounding area their protection has to be balanced with its economic cost. Therefore it is perfectly acceptable replacing an existing construction with a new one, always given that the style and characteristics of the original building is respected in order to protect the general feeling and atmosphere of the monumental district itself.

An example of heritage declared in the category of Historic District is the Minas de Reunion in the Sevillan town of Villanueva del Rio y Minas.[377] The site is a residential and industrial area developed between the 17th and 20th centuries for the extraction of coal in the left bank of the river Hueznar which has outstanding environmental, architectonic, anthropological and technological values, as well as spectacular landscapes. The Minas de Reunion possess some of the most outstanding example of workers living quarters in Andalusia, witness to the models developed in the 19th and early 20th centuries. The residential area is formed by the Barrio de la Confianza, built between 1896 and 1900 to substitute early slums, the Barrios de Velarde and la Constancia, built in the last decade of the 19th century, and the Barrios de Progresso, Cerro, Balbo and Transwall, which date from 1935. Also included in the Historic District is the Barrio de Casas Nuevas, which has probably the greatest artistic qualities, built in the 1920s as living quarters for company directors and highly qualified employees. Outstanding buildings in this zone include the Casa de Dirección, and the Casa del Ingienero Jefe. Other outstanding buildings include the church, built in 1927 and being the result of the transformation of an old warehouse, and the old school.

---

377 Decreto 35/2002, de 5 de febrero, por el que se declara Bien de Interés Cultural, con la categoría de Conjunto Histórico, Las Minas de la Reunión, sitas en Villanueva del Río y Minas (Sevilla).

The mining complex maintains the mark of the great technological advances in machinery, energy storage and systems of extraction made during the 19th century. The vast majority of these installations are found in buildings with interesting historic characteristics, with a predominance of usage of brick facing, although use of more modern architectural techniques involving the use of iron structures can be observed. The most outstanding buildings are situated near extraction shaft number 4, active between 1880 and 1908, being used later for storage.

The perimeter of the Historic District includes all the spaces, immovable objects and elements which made up the Minas de la Reunion throughout its history, as well as other areas closely related to the exploitation, including course of the Ribera de Hueznar, the bridge, the railway line and the storage ponds.

D) Historic Site

Article 15.4 defines historic site as "a place or natural landscape linked to events or memories of the past or to popular tradition, cultural or natural creations and works of man having historical, ethnological, paleontological or anthropological value".

An example of Bien de Interés Cultural declared in the category of Historic Site is the docks area on Valleseco beach in Santa Cruz de Tenerife. The group of docks, factories and warehouses on the Valleseco beach bear witness to the industrial heritage of the Archipelago, especially to the type of harbour area that grew with the expansion of the capitalist system linked to the first industrial revolution. The area is the only remaining stretch of the maritime area of Santa Cruz de Tenerife that is more or less conserved intact, preserving those elements representative of one the most important stages in the port's history, and in consequence, in Santa Cruz's history.

E) Anthropological Heritage

Scholarly opinion is divided as to what should be understood as "anthropological" (Latin countries use the term "etnográfico"). Although the study of man and his activities dates back to Aristotle, the 19th century sees the real beginnings of a systematic attempt to create a science studying the human species in its widest dimension. For Alvarez Alvarez

> anthropology as a science of culture or of the cultural heritage of primitive civilizations, has moved on to become a science that studies the cultural

> heritage of modern populations, strongly related to the spiritual sciences, such as History and Sociology, but also by the development of technology in the sense that these developments in production, tools or machinery form part of their culture.[378]

An interesting example of the regulation of anthropological heritage is found in the Heritage Law of the regional government of Cantabria, which states that,

> amongst the movable objects of anthropological heritage are found all those objects connected to the activities of the people of Cantabria, insofar as their models respond to technical developments which have their roots in the region...Also included in the category of anthropological heritage are those items of movable or immovable property that are connected to productive, technological and industrial activities in Cantabria, both in the past and present, insofar as they bear witness to the way of life of the people of Cantabria.[379]

3.2 Urban Planning Law and Historic Properties

In many cases historic buildings can only be understood in the context of the area in which they are situated, having defined through time a given space and related landscapes. By destroying the related landscapes, we destroy many of their associated values. For that reason, the involvement of urban planning law in historic property protection is vital.[380]

The protection of cultural heritage and urban planning and development laws have always been, and carry on being, two parallel public policy areas that, although theoretically independent, are intimately interconnected. Their relationship turns upon the tensions that are inherent to their peculiar sectarian perspectives. On the one hand, heritage law gives great importance to the conservation and improvement of cultural heritage, identifying items of heritage as items whose protection is in the public interest, while on the other hand urban planning and development law, while not negating the values of heritage, has the role of integrating that very same heritage in the

378 ALVAREZ ALVAREZ, J.L. Estudios sobre el Patrimonio Histórico Español, Civitas, Madrid, 1989, p. 85.

379 Article 97 of Law 11/1998 of the 13th of Octuber, of the Autonomous Community of Cantabria.

380 ALONSO IBAÑEZ, M.R. "*Patrimonio Industrial. Notas a su insatisfactoria protección jurídica*", Revista de Patrimonio Cultural y Derecho **3**, 1999, p. 257 y sigs.

territory in question, from a logical standpoint of rational usage of resources and land.[381]

Cultural spaces are, therefore, submerged in the problems created by the attempts to impose a rational use of territory, and it is therefore understandable that any attempt to effectively legislate the protection of historic heritage should attempt to condition urban planning law. Urban planning law and heritage law thus becomes a united and interconnected block of law that protects heritage from a double perspective: the protection of the object itself, and the protection of the area in which it is found, and which it has undoubtedly helped to form.

Scholars are unanimous in signalling that one of the most important elements of the Spanish Heritage Law is its effect on those laws that refer to zoning and urban planning, with the text of the law full of conditions placed on urban planning and development, in particular article 20, which states that:

1. "Declaration of a historical district or site or archaeological area as Bien de Interés Cultural shall entail the obligation for the local council or councils in which they are located to draw up a special Plan for the protection of the area involved by the declaration or another of the types of plan included in town planning legislation providing that under all circumstances it meets the requirements of this Law. Approval of this Plan will require a favourable report from the public authority responsible for the protection of the cultural property in question. A favourable report shall be understood to have been issued three months after presentation of the Plan. The obligatory nature of this Plan cannot be obviated based neither on the pre-existence of another plan that goes against the protection nor on the lack of prior existence of a general plan.
2. The Plan referred to in the above section shall establish for all public uses the order of priority for its implementation in buildings and spaces suitable for this purpose. It shall also cover possible areas for integral rehabilitation allowing recovery of the residential area and of appropriate economic activities. It shall also contain criteria relating to the preservation of façades and roofs and installations on them.

381 See in this sense, PAREJO ALFONSO, L. "*Urbanismo y Patrimonio Histórico*", Revista Patrimonio Cultural y Derecho **2**, 1998, p. 5.

3. Until final approval of this Plan, the granting of licences or the implementation of those granted prior to initiating the proceedings for declaring the historical unit or site or the archaeological area shall require a favourable resolution from the public authority responsible for the protection of the property involved and under no circumstances shall it be allowed to carry out new alignments, alter building potential nor divide or combine land.
4. Once final approval of the Plan referred to in this article has been granted, the Town Councils involved shall be entitled to directly authorise work to implement the approved plan affecting only buildings that are not monuments or historical gardens and are not included within their surrounding area, and the public authority responsible for enforcing this Law shall be informed of any authorisations or licences granted in a maximum period of ten days after they are granted. Any work carried out under licences that go against the approved Plan shall be illegal and the appropriate public authority shall order that they be reconstructed or demolished by the organisation that granted the licence in question, without prejudice to the provisions of town planning legislation regarding liability for infringements".

The article is fundamental to the protection of heritage in Spain, and has been the basis for similar previsions in regional law. The power to regulate urban planning laws in Spain corresponds to the regional governments[382] and so in this section we must make reference to Valencian urban planning laws.

Valencian regional law 6/1989,[383] on the organization of the territory of the Valencian community establishes as one of its objectives the conservation of historic, artistic and cultural heritage. It establishes the obligation that zoning plans should include the marking of those areas of cultural, historic and archaeological heritage, as well as a list of those areas that have districts or buildings that, due to their exceptional value, are eligible to be declared as Valencian cultural heritage.

Law 4/1992 of the 5th of July establishes restrictions on development in all those areas subject to protection under heritage law, as well as those areas that, while not subject to a specific regime of protection, do possess values and characteristics that make them

382 Sentence 61/1997 of the Spanish Constitutional Court.
383 Ley 6/1989, De 7 De Julio, de Ordenación del Territorio.

worthy of special protection, being therefore theoretically unnecessary that they be declared elements of heritage.

Valencia's urban planning law, the LRAU,[384] includes amongst its guiding principles the protection of architectural heritage, the environment and landscapes. According to article 3.2 those buildings which are situated next to buildings with historic, architectural, typical or traditional characteristics must harmonise with each other and with the area as a whole.

This law is the framework in which urban zoning plans must be developed, and establishes that these must include delimitations of one or more traditional historic districts. In these areas the indiscriminate development or substitution of buildings is expressly prohibited, and any modification, reform or rebuilding must be harmonised with the area's character. The LRAU also includes previsions for specialised plans which compliment or modify the general zoning plan. It establishes that they must, amongst other things, adopt measures to protect buildings with cultural or architectural value.

The Supreme Court has underlined the importance of local zoning laws and catalogues in its sentence of the 26th of October of 2006. In this case, a development was authorised by Badajoz Council in the area of the Alcazaba, a Bien de Interés Cultural. The area was subject to special protection in its zoning laws, drawn up in accordance with article 32 of the Heritage Law, which obliges the creation of a special catalogue to protect the surrounding area of a BIC. This catalogue established categories of protection for all the buildings in the area, some of which were affected by the construction of a faculty of library studies. The council argued that it had acted correctly given that it had followed the procedures established in articles 22 and 36 of the heritage law, having received a favourable report from the Comisión de Bienes Inmuebles del Patrimonio Histórico de Badajoz, and that therefore the special protection measures given to the heritage in question in the catalogue did not apply.

The court disagreed with the local council, pointing out that following the procedures indicated in articles 22 and 36, and obtaining a favourable report from the aforementioned commission, did not mean they could then ignore local zoning and planning laws. The court ordered the council to restore all the elements of heritage that had been affected by the works to their original state.

384 Ley 6/1994, de 15 de noviembre, de la Generalitat Valenciana, Reguladora de la Actividad Urbanística, [DOGV núm. 2.394, de 24 de noviembre] modified by Law 10/1998 of 28th December 1998, Law 16/2003 of 17th December 2003 and Law 4/2004 of 30th June 2004.

3.3 Heritage of Local Importance in Valencian Regional Law

Valencian Heritage Law, developing State Heritage Law, includes all heritage in the Valencian community into the category of Valencian cultural heritage. This then has three subcategories: Heritage declared as Bienes de Interés Cultural; heritage that has some of the qualities that would be required for declaration as a BIC, and which make it worthy of protection; and finally, a catch all category: heritage not included in the previous two categories. Immovable goods included in the second category are called "Heritage of Local Importance". The 2007 modification of Valencia's regional law states that local Heritage must be declared in one of the following categories:

- Monuments of Local Interest;
- Traditional Historic Centre;
- Historic Garden of Local Interest;
- Ethnological area of Local Interest;
- Historic Site of Local Interest;
- Archaeological protection area;
- Paleontological protection area.

The new text also automatically recognises as Heritage of Local Importance the following specific types of building:

- Traditional Historic Centres, named as such in zoning plans;
- Industrial brick chimneys built before 1940;
- Windmills;
- Traditional buildings in the area of "Horta de Valencia";
- Docks built before the 19th century;
- Religious architecture previous to 1940;
- Exterior ceramic panels placed before 1940;
- Those items surrounding a Bien de Interés Cultural, determined as worthy of protection in the relevant decree.

The introduction of the legal figure of "Heritage of Local Importance" gives an important role to local councils in the protection of heritage. Local councils are expected to participate actively in the inclusion of heritage in the General Inventory of Valencian Cultural

Heritage through the modification of their Local Catalogues of Protected Goods and Spaces.

Law 4/1998 on Valencia's cultural heritage defines the idea of Heritage of Local Importance as those buildings that, although they do not posses those values that would merit declaration as a Bien de Interés Cultural, do indeed have a certain level of historic, artistic, archaeological, architectonic, palaeontological or anthropological significance. The law begins with the assumption that examples of Heritage of Local Importance exist in all local council areas, and therefore requires that, where the Catalogue of Protected Goods and Spaces, which has to be published within one year of the passing of the law, does not contain any reference to Heritage of Local Importance the council must explain the reasons for their absence.

The law therefore establishes a *iuris tantum* presumption in favour of the existence of examples of Heritage of Local Importance, and places the burden of negative proof upon the local council. This initiative in the law is to be applauded, as it forces local councils to act and to investigate what heritage exists in their area of influence, raising local awareness and preserving many buildings that would otherwise have passed through the safety net provided by the law.

Before a council includes a building considered Heritage of Local importance in the general Valencian register, it must first include it in its local catalogue. The catalogue and its modifications are subject to approval by the Conselleria of Culture, Education and Sport. The approval by the Conselleria is obligatory and binding, both in terms of whether the catalogue correctly includes examples of Heritage of Local Importance and in terms of the actual level of protection to be given to the buildings included.

The protection enjoyed by Heritage of Local Importance is immediately applicable after the opening of the period of public audience during the administrative procedure to establish the catalogue. From this moment on, the Heritage of Local Importance is subject to the dispositions relative to the regime generally applicable to catalogued heritage, included in Law 4/1998, to the regime included in Urbanism Law, and to any additional protection which is included in the catalogue itself.

Any project for building work on an example of Heritage of Local Importance must include a study of the historic values associated with the building, its actual state and any deficiencies present in the building. It must also detail the work proposed and the effect it will have on the historic values of the building. The project will have to be written by a competent expert in each field affected by the project. Within one month after the finalization of the work, the projects promoter must present a description of the work carried out to the council that conceded the licence, and this must forward it to the Conselleria of Culture. All licences issued by the local council for work on Heritage of Local Importance must be communicated to the General Directorate of Linguistic and Valencian Cultural policy within 10 days of issuance.

In those cases where Heritage of Local Importance is to be declared by the council to be in a state of ruin, the Conselleria of Culture must be directly involved in the administrative procedure, and must therefore be notified before any action is taken. The administrative procedure must be open to public audience for a period of at least one month. The state of ruin of a building declared Heritage of Local Importance which has come about as a direct result of non compliance with the legal dispositions included in Law 4/1998, will oblige the owner to carry out all necessary repairs to return the building to its original state.

3.4 The Legal Obligation to Conserve Buildings in Spanish Heritage Law, and Limits on Reconstruction

The conservation and maintenance of all buildings is an obligation which affects all property owners. This legal obligation to conserve buildings is included in national and regional legislation. In Valencian law, it is laid out in article 86 of the LRAU, and is especially intense where the property involved forms part of cultural heritage. This increased obligation has its origins in the social function of heritage and its consideration as part of the general public interest, as well as the necessity to guarantee public access[385] to those special values that are inherent in the structures involved.[386]

Article 36 of the Spanish Heritage Law also establishes the general principle that those items which are considered as Spanish heritage should be conserved, maintained and protected by their owners or users. Where this duty of care is not fulfilled, public authorities can carry out the conservation or maintenance work on behalf of the above mentioned stake holders, with these being responsible for any costs

385 Public access is not limited to a simple contemplation of the outside of buildings, but is of a more personal nature, and is a direct application of the right to access to culture defined in article 44.1 of the Spanish Constitution.

386 See in this sense, ABAD LICERAS, J.M. La situación de ruina y demolición de inmuebles del Patrimonio Histórico, Montecorvo, Madrid, 2000, p. 37.

incurred. Abad Liceras argues that the obligation to conserve in Spanish heritage law converts the private owner of heritage into a mere possessor of the right of usufruct, that is, the right to use and enjoy the profits of the heritage so long as it is in no way deteriorated or misuse with the true owner of the heritage being, in a sense, society as a whole.

Bienes de Interés Cultural can be reconstructed and repaired under Spanish heritage law, subject to permit. The LPHE[387] includes a set of measures designed to limit intervention on items of cultural heritage, stating that public authorities must give priority to the conservation, consolidation and improvement of Bienes de Interés Cultural. Restoration work should tend towards repair and consolidation, avoiding any reconstruction except where original materials are used. When non original materials are used and those extra structures which are added for reasons of stability, they must all be easily recognisable, avoiding any possibility of confusion between what is original and what has been added.

Any restoration must respect those elements added in all previous times, eliminating them only when they suppose an evident degradation of the property or where their elimination will allow for improved understanding of the property. Those parts that are eliminated must be registered and documented.[388]

The limits to reconstruction are highlighted in STS of the 16th of October 2000 referring to the reconstruction of the Roman Theatre in Sagunto by the Valencian government, declared a National Monument in 1986. The theatre was remodelled extensively in 1988, under a project designed by two renowned architects, Manuel Portaceli and Giorgio Grassi. Marble was used to cover the original seating, and extensive additions being made to the structure of the theatre. A report issued by the Academia de Bellas Artes de San Carlos declared that the project created, using new materials, a "theoretical idea" of a roman theatre, and that the use of new material ran the risk of destroying the very essence of the monument. A local lawyer, Juan Marco Molines sought judicial review of the administrative action, alleging that the restoration work violated Spanish Heritage Law.

The Spanish Supreme Court agreed with the plaintiff, pointing out that the LPHE establishes a positive limit – that interventions in monuments are designed to conserve, consolidate and rehabilitate –

387 Article 39.

388 Article 39.3.

and another negative – avoid reconstruction. This negative restriction does have an exception – that original parts are used and that their authenticity can be proved. The court felt that this legal restriction should be interpreted as narrowly as possible, not only because of the principles that apply to works of conservation, but also because of the laws evident opposition to reconstruction of monuments.

> One shouldn't reach the conclusion that the existence of small amounts of original remains will allow reconstruction of a monument, even when perfectly documented. What the law permits is the reconstruction of those parts that can be rebuilt with original materials, and whose authenticity can be proved. For example, a wall that has fallen down can be rebuilt using the original stones.[389]

Cultural, artistic or architectural considerations are irrelevant, as is any international recognition of the project as such.

According to the court, the reconstruction of the Roman Theatre is so far reaching that it cannot be considered simply as a work of conservation or repair, but is rather a full scale reconstruction, and in some cases, a reconstruction of what appeared to be a "generic" Roman Theatre, with the original form and style of many of the original elements being completely ignored. While the qualities of the reconstruction may be considered artistically or architecturally correct,

Roman Theatre, Sagunto © Julian Negredo Sanchez.

389 FJ IV STS 16th of October.

they do not follow the criteria established in the LPHE, to which public authorities should be bound.

The court contrasts this with the case law that has been established over the discretionary privileges of public authorities. The court considers that the public authority does have discretion to decide between various solutions that are juridically indifferent, but when one of the solutions is contrary to law, then the court is not invading administrative discretion when annulling an action, but carrying out its role of judicial review of administrative action.

The court points out that

> had the legislator impregnated the law with artistic and architectural criteria for the reconstruction of monuments, then the public authority would be free to choose between all the different alternatives proposed, and that the courts would have no jurisdiction to challenge that discretionary decision, based on artistic, cultural or architectural considerations. This is not the case here, and the court correctly declares that the public authority has overstepped its discretionary powers and that it should have been opposed to the construction that is contrary to law.

Some authors have pointed out that this interpretation of reconstruction under the law is, at times, too restrictive. Martin-Retortillo feels that this sentence leaves a bitter taste and does not seem to appreciate “the changing nature of historic monuments”. While the author does accept that this approach will avoid many disastrous reconstructions, he feels it also “introduces a rigidity and irrationality that can be extremely harmful in some occasions”. It is somewhat ironic that the main level of protection available to immovable heritage under national law is so restrictive that public authorities, on occasions, may be reluctant to declare a monument, given that the declaration will severely restrict any future development on the site, and condemn it to immobility.

Valencian regional law, as a reaction to the Supreme Court’s sentence, which is considered too restrictive, lays out new criteria to be used in the reconstruction of Monuments. In a modification to the law published in 2007, article 38 is modified to read as such:

> Any intervention on a Monument declared as a Bien de Interés Cultural should be aimed at preserving and increasing those cultural qualities which make up the monument, and which follow the following criteria:
>
> a) The intervention will respect the characteristics and essential values of the property. Its volumetric, spatial

and artistic properties will be conserved, as will those additions in latter years that have enriched its original values. Any suppression that is authorised will be fully documented;

b) The integrity of the property will be preserved and any separation of any of its essential parts will be authorised...;

c) Bienes de Interés Cultural are inseparable from their surrounding area. Their movement will only be authorized where it is unavoidable in the public interest, and must be authorized by the department of culture;

d) Total or partial reconstruction of the property may be authorized where some original elements remain, or where documentary evidence of what is missing exists. The reconstruction should attempt to use, where possible, the use of original techniques and materials. The result should be made understandable through the use of drawings, models, virtual presentations or whatever other technique allows people to differentiate between original and added elements.

The new redaction of Valencia's regional law would therefore have permitted the reconstruction reversed by the Supreme Court, requiring the use of original materials only "where possible", and making the requirements for differentiation of original and new elements more flexible.

3.5 Financial Tools in Spanish Heritage Law for the Protection of Historical Property

3.5.1 Fund for Protection and Rehabilitation of Historic Property

Spanish Heritage Law, fulfilling the constitutional mandate to promote heritage conservation, includes various measures designed to finance historical preservation.

Article 68 of the law lays out that any public works financed by the national government should have one percent of their budget allocated to the financing of work on the preservation or enrichment of the Spanish Historical Heritage or for promoting artistic creativity, preferably on the actual site of the work or in the immediate surroundings. If those works are done by private companies due to an administrative concession, one percent of the total budget should be used, even where the national government is not financing the works.

Some public works are exempted:

- Those for which the total budget does not exceed 600,000 Euros;
- Those relating to the security and defence of the State and the safety of public services;
- Those public works that are designed to protect heritage.[390]

The public administration responsible for the works can choose between using the funds to finance projects in the area of the works, or on any other Bienes de Interés Cultural that they administer, or including the funds in the yearly plan for protecting and enriching heritage, which is financed by the funds raised by this public works levy.

Regional laws include similar previsions for the rehabilitation of heritage. Valencian regional law, for example, in its article 93, establishes that at least 1% of the total destined to public works in the previous year, should be dedicated to programmes of investigation, preservation, promotion and restoration of Valencia's heritage, all of which should be organized by the department of culture. The law requires that, to this end, the relevant department should elaborate an annual Plan for the Conservation and Enrichment of Valencia's Cultural Heritage, which must lay out clearly its annual objectives and the destination of the funds available.

3.5.2 Tax Incentives for the Protection of Historic Properties

Article 70 of the heritage law allows a deduction on tax liability equivalent to 20% of the amount spent on purchase, preservation, repair, restoration, dissemination and exhibition of Bienes de Interés Cultural, with a limit of 30% of each individuals total tax liability. This deduction also applies to donations made for the preservation of Spain's historical heritage. Cultural Heritage is also exempt from local property taxes where owners have undertaken works of conservation and rehabilitation.

3.5.3 Preferential Right of Purchase

Article 38 of Spain's heritage law gives the state a preferential right of purchase on property that has been declared as a Bien de Interés

390 Article 58.1 RD 111/1986.

Cultural. According to this provision, when a private sale is agreed of a BIC or on any other property included in the register, the owner must notify public authorities, declaring the price and conditions proposed for the sale. This condition also applies to auction houses where they are planning to sell any item forming part of Spain's historic heritage.

The State then has the right to purchase the property at the same price agreed for the sale, or for the price reached at auction. The payment must be made within two financial years, irrespective of the conditions agreed in the original sale. The state may however agree other terms of payment. The state must notify its intention to exercise this right of purchase within two months of receiving the notification. Where proper notification is not given, the time period is extended to six months from the moment that reliable information on the sale is received.

Regional administrations may also exercise these very same rights. However, the State has the first option, provided that such property is acquired for a museum, archive or State-owned library.

This right has been reaffirmed in Supreme Court sentence of the 21st of November 2006. In this case, the plaintiffs were owners of Goya's painting "The Countess of Chinchón", declared as a BIC. The owners informed the administration of their intention to sell the painting for 24,000,000 Euros. The agreed payment terms were for cash to be paid 30 days after sale. The State decided to pre-empt the sale according to the rights given in article 38 of the law, and so received delivery of the painting in the Museo del Prado. Seven months after delivery, the owners still had not received payment, and sued the public authority for the legal interest on the money owing. The plaintiffs alleged that the right to pre-empt a sale implies that the public authority must fulfil all the conditions agreed in the original sale, and that therefore payment should have been made in 30 days. If this were not so, article 38 would establish a right in favour of public authorities that seriously prejudiced the property rights of the owner of the Bien de Interés Cultural. The public authority, for its part, argued that the law is clear in its intentions: the state pre-empts the sale and has the duty to pay the same price agreed, but that the terms of payment are decided freely by them, with the only limit being two financial years.

The court agreed with the State's interpretation of the law, pointing out that the payment of interest would be obligatory six months after the end of the two year time period. The court is happy that in this case the property rights of the individual have been fully respected. The decision to put the item up for sale was taken freely by the owners, and the public authority acted correctly in delaying payment for up to two years.

4 The Sagunto Steel Works

The following case study relates to a group of properties that have many of the features that would permit its declaration as a historic district within Spanish state or regional heritage law. The elements that go to make up the area are actively promoted as industrial heritage by the local tourist board, but, nevertheless, the site enjoys no specific protection under heritage law against redevelopment. In fact, the general zoning plan allows for the construction of up to 200 housing units on the site. The problems faced by the area illustrate the difficulties of protecting heritage within a framework which depends on administrative action and where there is an evident conflict of interests between the sites owners and those interested in the preservation of the site. This difficulty is especially evident in the case of disperse industrial heritage, where the elements that give the site specific interest are not in one, easily located location, but are rather dispersed throughout an urban area, and in many cases, in varied states of decay.

Roman Castle, Sagunto © Julian Negredo Sanchez

Sagunto is a town situated on the east coast of Spain, 20 kilometres north of Valencia. It has an incredibly rich collection of buildings that represent Spain's classical heritage, including many roman buildings such as the Temple of Diana and a Roman Theatre, as well as a well conserved Jew's Quarter and a quite breathtaking castle. It also possesses an important industrial legacy, which is worthy of study as a fine example of an industrial "factory city".

Sagunto's industrial area has a history that stretches back over a hundred years, and is a major protagonist in an important period in industrial production which finalizes with one of the most traumatic post-industrial conversions in recent Spanish history. The industrial district known as the "Gerencia" is probably the most important architectural and social reference remaining of all the buildings, districts, infrastructures and machinery connected to the social development policy of the steel industry of the early 20th century.

In the area included in what is known as the "Gerencia" the following elements can be found: the church of our lady of the Begoña, the launderette, the Altos Hornos del Mediterraneo cooperative, the casino, the gardens and the management offices and the management quarters in the "garden city". The surrounding area also contains interesting elements such as the old apprentice's warehouse, the Alameda, the spare parts warehouse, the worker's living quarter and various "groups" of living quarters. All these buildings are surrounded by a historic garden which, due to the nature of the species found there and its design denotes a planning for which historic documentation exists.

Main Entrance to the Gerencia, Puerto de Sagunto © Julian Negredo Sanchez.

Entrance door to offices of AHV. Puerto Sagunto © Julian Negredo Sanchez.

The historic district of the Gerencia is contained in a perfectly delimited area of around 60,000m^2, and although its historic and urban character cannot be understood without reference to the surrounding areas, such as the Sagunto Port area, the old hospital, the sanatorium, the abattoir, the market, the old cinema and the sports stadium, it is without doubt the most culturally relevant area and the most easily identifiable.

As previously seen, for a property or set of properties to be included as a Bien de Interés Cultural under Spanish and Valencian heritage law, two factors must be present: a unique quality and cultural relevance. The following text will examine these two elements with reference to the Gerencia in Sagunto.

As a starting point it is worth describing the history of Sagunto's industrial growth. Sagunto was, at the beginning of the 20th century, an almost entirely agricultural town, an almost entirely grape growing area which was beginning to transform over to orange growing.

Head offices, Ojos Negros. Courtesy of Miguel Angel Piqueras. Redigitalized by Julian Negredo.

Between 1900 and 1907 a railway line was built between the mines at Ojos Negros[391] in Teruel, rich in iron deposits, and Sagunto and special docks were built in Sagunto to transport iron to the rest of Europe.

391 The history of the mining area of Ojos Negros is firmly connected to to the history of Sagunto. The Compañia of Sierra Menera was founded by Ramón de la Sota, a Basque business man, who became interested in the area as Basque deposits of iron began to be over mined. Sota owned at one point over 25 ships to transport minerals to Europe, and had his own shipbuilding company in Bilbao (Talleres Euskalduna), his own insurance company (Seguros el Polar) and his own bank (Banco del Comercio).

The mining area of Ojos Negros, about 200 kilometres from the sea, is formed by 22 iron ore mines in an area that extends over approximately 1500 hectares. The iron extracted was of a type much in demand, especially in the United Kingdom. In the commercial document where Sota announced his intention to start mining in Ojos Negros, he also announced his intention to go into the steel business. He therefore controlled all parts of the production chain, from the mine, through to the railway and the steelmaking facilities, installing high ovens for the transformation of minerals into laminates.

Up until the early 1920s, the mineral extraction in Ojos Negros was carried out without machinery. In 1921 the company purchased two frontal steel diggers. In 1932 production was halted due to the great depression, and restarted only in 1941, when the raw material became essential in the reconstruction of Spain following the end of the Spanish Civil War.

The need for raw material from the steel works in Sagunto grew incessantly in the late 1940s and in 1950 the mining company purchased new diggers powered by diesel fuel. In 1969 a transportation belt of over 2.5 kilometres was installed in the Montiel tunnel, which allowed for a considerable increase in production levels. Around this very same time, a new plant for classifying the different materials extracted was built.

In 1972, the company signed an agreement with the Spanish state railway system for the use of their infrastructure, which meant a loss of almost 350 jobs at the site. The company finally closed in 1987, after the demise of the parent company in Sagunto. At present, several buildings remain on the site, including the classification warehouse, the old maintenance warehouse for heavy vehicles, the two railway workshops and part of the original railway line.

Miner's train, Ojos Negros. © Julian Negredo Sanchez.

CMSM railway with mines in background, Ojos Negros. Courtesy of Miguel Angel Piqueras. Redigitalized by Julian Negredo

Authors have pointed out the "ex-novo" characteristics of the industrial nucleus of Sagunto. It is a perfect example of an industrial city built from scratch for strictly industrial reasons, and of one that is completely isolated from the original infrastructure that surrounds it. As all industrial cities of these characteristics, it enjoyed a hugely dependant relationship with the industrial process that it was born to support.[392] The new city was built with astonishing speed. A contemporary travel writer wrote:

> What a spectacular view awaits us. Are we by any chance dreaming? Is this Sagunto's port, which we have always known by its four closed warehouses, or is it a huge city that has sprung from the ground as if by art of magic, with great buildings, workshops, factories, ovens, electrical installations, railway, docks and huge numbers of people? Only six years ago only a few rickety boats would stop at this port, and now sirens announce the arrivals and departures of huge steam boats! The forty million pesetas of capital invested by the Compañia Minera de Sierra Menera has created a miracle. An artificial city has risen as if a toy. What is at the moment inappropriately called Canet Port, is truly Sagunto's Port, a new Hannibal that will, sooner or later, finish with the remnants of this heroic city.[393]

In 1923 the company *Compañia Siderúrgica del Mediterraneo* was formed, with the aim of producing iron from the deposits in Ojos Negros. The factory had 1800 workers with people coming from all over the Spanish peninsula to work in Sagunto. The company went into crisis in the 1930s, during the great depression, and was absorbed into the company *Altos Hornos de Vizcaya* in 1940, and was finally taken over by the company *Altos Hornos del Mediterraneo* in 1974. The integrated steelworks were finally closed in 1984 with more than 2000 workers being laid off or given early retirement, creating a huge economic and social crisis in the area of Sagunto.

The traumatic crisis created by the shut-down of the steel plant, and the need to move on with their lives, meant that the very last thing that the people of Sagunto were worried about in the immediate aftermath of the closure was the preservation of the installations of the old steelworks. At this given point of time the preservation of the *Gerencia* for the cultural significance of the old installations must have seemed to be one of the least important problems faced by the population.

Indeed, one author points out that:

> after the closure of the factory, the Saguntine people were faced with the dilemma of what to do with the vestiges of their industrial past. Many people opted to turn their backs and forget that which had formed part of their history. Nevertheless, scarcely a decade after the closure of *Altos Hornos del Mediterraneo*, the same city which had previously tied its future to the industrial activity of steel making, has found in the old installations a new way

392 See RODRIGUEZ CUADROS, E. "Los hijos de Promoteo, historia de un suelo amado" in Reconversion y Revolución, op. cít. p. 25.

393 MARTÍNEZ ALOY, J. "Provincia de Valencia", en CARRERAS, *Geografia general del Reino de Valencia,* Alberto Martín, Barcelona pp. 50–51.

> of understanding and uniting their city as an enclave of industrial heritage, and has made a great effort to five new meaning and value to the industrial remains that were left after the trauma of 1984.[394]

Economic crisis and industrial reconversion not only give rise to the question of what to do with large, obsolete, industrial areas, but also what to do with a large amount of machinery and related buildings. As Aguilar Civera points out, evidently we cannot preserve the industrial building whilst discarding those elements that are closely related to the production process – railways, canals, workers quarters etc. The existence of each element is difficult to explain without reference to the others.[395]

Perhaps a necessary element in deciding whether something bears cultural witness to a recent époque is the emotional impact it has on the inhabitants of the area, or, in other words, the emotional reactions provoked by the heritage. There is no doubt that the industrial heritage of Sagunto is deeply ingrained into the emotional psyche of the population of the area. Rodriguez Cantos eloquently describes what the area means to many Saguntinos:

> those of us who lived in the Port of Sagunto and learnt the name of the winds by the different sounds they made in the factory, heard at night or at sunset, are still deeply affected by this memory: not only that, but the smell of the fallen leaves from the huge eucalyptus trees of the Fornas garden, or of the gold dust from the minerals drifting along the road to the docks. It pulls at the emotions just as much as the sight of the giant industrial buildings and factories, all built with such functional aesthetics. The warehouses and the brick structures of the coking ovens are perfect examples of how beauty can be found in the most unexpected places. One cold day in January I had the opportunity to contemplate this landscape from the second platform of High Oven number 2. I cannot deny that it was a turning point in my emotional and intellectual life. There was the sea, so close, the land, mixed with ash and ice, disturbed by the disappearance of the chimneys. I lived, in a few seconds of conscience, a sense of loss similar to all those who find in these feelings reasons to save what Juan Gil Albert called 'pieces of that loved black land'. There are times when the passing of history hurts us deeply. That day, from the heights of the last Prometheus of Sagunto's steel industry, I thought more than ever that we are but 'little men on giant's shoulders'.[396]

---

394 REVERT ROLDAN, X. & MONTIEL ROIG, G. Presentación al libro Reconversión y Revolución, Industrialización y Patrimonio, edit. Universidad de Valencia, p. 15.

395 AGUILAR CIVERA, I. "*La investigación del Patrimonio Histórico industrial del Puerto de Sagunto*" in Reconversión y Revolución, Industrialización y Patrimonio, Universidad de Valencia, Valencia, 1999, p. 37.

396 RODRÍGUEZ CUADROS, E. "Los hijos de Prometeo: historia de un negro suelo amado", op. cít., pp. 26–29.

High Oven number 2. © Julian Negredo Sanchez.

Night view of High Oven 2 and surrounding area. © Julian Negredo Sanchez.

View of High Oven number 2 from the offices of AHV, Gerencia, Puerto de Sagunto. © Julian Negredo Sanchez.

Aside from the emotional attachment to our recent industrial heritage, there are many reasons for considering the Gerencia bears witness to our recent history, in an exceptional and unique way. One of the most important elements in the development of Sagunto's steel industry was the influence of contemporary thinking on urban planning, and especially from the so called "groups of living quarters". In Sagunto, some 14 different groups have been catalogued, with a total of 1650 living quarters, which make of 14% of the total living space in Sagunto in the present day. There is no other place in the Valencian regional community that has such a high concentration of social living quarters.[397] These groups were promoted mostly by the steel company itself, thus substituting official administration bodies such as the *Obra Social del Hogar* and *the Instituto Nacional de la Vivienda.* These groups are not only unique in the Valencian community, but they also are an integral part of the urban layout of Sagunto. They constitute a true witness to the social development of the city and the life of its people, and in the opinion of many, best represent what it means to have lived and grown up in the city.

397 ALONSO DE ARMIÑO, L. "La excepcionalidad de la vivienda social y obrera en el Puerto de Sagunto" op. cít., p. 74.

Deteriorated state of the Cooperativa de Productores, Puerto de Sagunto.
© Julian Negredo Sanchez.

Engineers Chalets, Gerencia, Puerto de Sagunto. © Julian Negredo Sanchez.

A perfect example of these groups is the *ciudad jardín de los ingenieros*, a segment of the town which has its own morphological identity in the confines of the factory itself. It is a minuscule garden city built in the 1920s made up of around 20 houses, surrounded by private gardens, wooded streets and avenues, a square, play areas for children and the management offices of the steel making company.

Sagunto is also an important example of urban planning, all the more valuable as the process of industrial reconversion makes these models continually scarcer. Luis Piñon points out that the urban area of the port of Sagunto is one of the richest in our community insofar as it constitutes one of the purest models of the “worker’s city”, reuniting in one area spaces dedicated to work, management, leisure and living. The absence of other examples of this type of urban planning makes those that we do possess all the more valuable and worthy of conservation. This perspective “allows us to speak, not only from a certain value of antiquity, but also from the perspective of the cultural witness to the definition and theorization of the industrial city”.[398]

On the 30th of April 1998, a popular initiative was presented to the Dirección General de Patrimonio Cultural de la Generalitat Valenciana, requesting that the various elements of industrial heritage in Sagunto Port be declared as a Historic District. The initiative included abundant documentation on the social and historic importance of the site, and enjoyed wide support amongst political parties and social groups in the area. The initiative was however not accepted by the relevant public authority.

The public authorities then put forward in the year 2000 a plan to create a leisure area dedicated to the scenic arts, which would make use of many of the elements that have been studied here. The plan, supported by the same groups that at first promoted the declaration as Bien de Interés Cultural, implied passing the property to public ownership. This however required agreement between public authorities and the owners of the site, in order to compensate the latter for loss of development rights on the land given over to public ownership. At the date of writing no such agreement has been reached.

The situation described here illustrates one of the shortcomings in heritage protection laws. The sites described here have great potential in two different areas, In the first place they enjoy great economic potential as real estate, and the owners of the site are understandably

398 LUIS PIÑON, J. “Patrimonio Histórico y Proyecto Urbano: el caso de Port de Sagunt”. op. cít. p. 70.

reluctant to forgo the economic benefits of redevelopment. In second place, the site enjoys great cultural potential as a site given over to public use. The proposed solution, an area dedicated to scenic arts, requires that the land be transferred to public ownership, and this of course requires both parties to be in agreement on the economic conditions involved.

Declaration of the site as a Bien de Interés Cultural would have supposed a protection possibly too rigid for the areas involved, especially taking into account the previously commented Supreme Court decision relating to Sagunto's Roman Theatre, which effectively limits rehabilitation on the site. The refusal of the public authority to accept the popular initiative to declare the properties as a Historic District effectively leaves the area unprotected, and meanwhile, many of the buildings, including the area of the Gerencia itself, are in a state of grave decline. The site illustrates the need for more imaginative models for protection of some types of industrial heritage.

Of the three systems studied in this publication, the one that seems most comfortable with the protection of industrial heritage is the legal framework existing in the United Kingdom. Industrial sites have been protected adequately using the techniques of Conservation area creation, listed buildings, and scheduled monuments. The flexibility of the listing system, with the existence of three grades of protection, also facilitates the process. The enormous number of buildings that are listed or scheduled and the large numbers of urban areas that have been designated as Conservation area is a testimony to the effectiveness of the system. A useful model that would serve to protect Sagunto's industrial heritage is the Stroud Valley Industrial Conservation Area, which as seen previously takes in elements of industrial heritage spread throughout the town, and which takes into account that that "many buildings that lack an obviously 'industrial' appearance may actually have links to the industrial heritage of the area: mill workers' cottages, for example, or buildings and structures that relate to the evolving transport infrastructure which was so crucial to the Stroud valley's industrial progress". The system is again useful in both offering protection to individual historic buildings and areas and in utilising the resources protected as an important offer of industrial tourism.

The Spanish system offers a high grade of protection to buildings that are declared as heritage. Nevertheless, the lack of flexibility in the level of protection offered can make the system too rigid. Additionally, the designation system is cumbersome and can take a large amount of time and resources, which may make public administrations wary of overusing the tool. Regional laws in Spain do offer more flexible systems, and urban planning laws do offer a certain role to local governments in deciding those buildings and areas that are worthy of protection. Generally however, decision making on what areas are to be protected is centralised in regional governments. It is however in this country that the Constitutional theory of the social function inherent to property ownership is most fully developed.

The American system, as we have seen, does not embrace theories analogous to the theory of the social function of property, although courts do accept public interest arguments in some cases. This has allowed the development of local protection of historic properties in the case of local historic preservation ordinances that in many cases offer effective protection to industrial property. However, the economic interests that are linked to very large areas of real estate can sometimes make protection of this sort of property difficult. The type of voluntary systems of cooperation between public administrations and local stakeholders through grant programmes and techniques such as National Heritage Areas can be very effective in the protection of this type of historic property. One useful model is the Rivers of Steel Heritage Area, which, as seen in Chapter 2, promotes the steel heritage in the South West of Pennsylvania, including the Pittsburgh Area, which, in the 10 years since its foundation, has been involved in various projects to protect historic resources, to maintain cultural traditions, to promote economic development and to create educational resources to foment interest in the historic heritage represented by the area.

As we have seen in the case of Sagunto's Gerencia area, the protection of industrial historic properties can involve many conceptual difficulties. In many cases the significance of the area is difficult to see without recourse to expert academic advice. While it is, for example, easy to see the significance of a Roman castle, it is sometimes difficult to see the significance of a certain type of living quarter that may demonstrate an approach to urban planning that was radical at the time. The case of Sagunto also illustrates how, when an industry closes, it may be the area itself that wants to just move on, demolishing buildings and placing new industries or activities in their place. In these cases, and especially where strong economic interests are involved, it can be difficult to persuade public administrations to bring down the full force of the law's protection on an industrial site, especially if, once protected, the law is inflexible on the uses and types of reconstruction that can be carried out.

International efforts to protect world heritage depend on the willingness of governments to nominate sites, and to take on the responsibilities that acceptance on to the World Heritage List brings. While in the recent past there has been a growth in industrial heritage on the list, the large majority of sites are included in traditional conceptions of what should be considered heritage. Our study has shown that there is increased acceptance at international, national and local level that it is not only traditional historic properties that find a place in legal systems designed to protect buildings that serve as witness to our past, but also industrial buildings and sites that have served to define perhaps the most important periods in human history and which this author believes deserve our recognition and protection.

Appendix 1

Convention Concerning the Protection of the World Cultural and Natural Heritage

Adopted by the General Conference at its seventeenth session Paris, 16 November 1972

The General Conference of the United Nations Educational, Scientific and Cultural Organization meeting in Paris from 17 October to 21 November 1972, at its seventeenth session,

Noting that the cultural heritage and the natural heritage are increasingly threatened with destruction not only by the traditional causes of decay, but also by changing social and economic conditions which aggravate the situation with even more formidable phenomena of damage or destruction,

Considering that deterioration or disappearance of any item of the cultural or natural heritage constitutes a harmful impoverishment of the heritage of all the nations of the world,

Considering that protection of this heritage at the national level often remains incomplete because of the scale of the resources which it requires and of the insufficient economic, scientific, and technological resources of the country where the property to be protected is situated,

Recalling that the Constitution of the Organization provides that it will maintain, increase, and diffuse knowledge by assuring the conservation and protection of the world's heritage, and recommending to the nations concerned the necessary international conventions,

Considering that the existing international conventions, recommendations and resolutions concerning cultural and natural property demonstrate the importance, for all the peoples of the world, of safeguarding this unique and irreplaceable property, to whatever people it may belong,

Considering that parts of the cultural or natural heritage are of outstanding interest and therefore need to be preserved as part of the world heritage of mankind as a whole,

Considering that, in view of the magnitude and gravity of the new dangers threatening them, it is incumbent on the international community as a whole to participate in the protection of the cultural and natural heritage of outstanding universal value, by the granting of collective assistance which, although not taking the place of action by the State concerned, will serve as an efficient complement thereto,

Considering that it is essential for this purpose to adopt new provisions in the form of a convention establishing an effective system of collective protection of the cultural and natural heritage of outstanding universal value, organized on a permanent basis and in accordance with modern scientific methods,

Having decided, at its sixteenth session, that this question should be made the subject of an international convention,

Adopts this sixteenth day of November 1972 this Convention.

I DEFINITION OF THE CULTURAL AND NATURAL HERITAGE

Article 1

For the purpose of this Convention, the following shall be considered as "cultural heritage":

monuments: architectural works, works of monumental sculpture and painting, elements or structures of an archaeological nature, inscriptions, cave dwellings and combinations of features, which are of outstanding universal value from the point of view of history, art or science;

groups of buildings: groups of separate or connected buildings which, because of their architecture, their homogeneity or their place in the landscape, are of outstanding universal value from the point of view of history, art or science;

sites: works of man or the combined works of nature and man, and areas including archaeological sites which are of outstanding universal value from the historical, aesthetic, ethnological or anthropological point of view.

Article 2

For the purposes of this Convention, the following shall be considered as "natural heritage":

natural features consisting of physical and biological formations or groups of such formations, which are of outstanding universal value from the aesthetic or scientific point of view;

geological and physiographical formations and precisely delineated areas which constitute the habitat of threatened species of animals and plants of outstanding universal value from the point of view of science or conservation;

natural sites or precisely delineated natural areas of outstanding universal value from the point of view of science, conservation or natural beauty.

Article 3
It is for each State Party to this Convention to identify and delineate the different properties situated on its territory mentioned in Articles 1 and 2 above.

II NATIONAL PROTECTION AND INTERNATIONAL PROTECTION OF THE CULTURAL AND NATURAL HERITAGE

Article 4
Each State Party to this Convention recognizes that the duty of ensuring the identification, protection, conservation, presentation and transmission to future generations of the cultural and natural heritage referred to in Articles 1 and 2 and situated on its territory, belongs primarily to that State. It will do all it can to this end, to the utmost of its own resources and, where appropriate, with any international assistance and co-operation, in particular, financial, artistic, scientific and technical, which it may be able to obtain.

Article 5
To ensure that effective and active measures are taken for the protection, conservation and presentation of the cultural and natural heritage situated on its territory, each State Party to this Convention shall endeavour, in so far as possible, and as appropriate for each country:

a) to adopt a general policy which aims to give the cultural and natural heritage a function in the life of the community and to integrate the protection of that heritage into comprehensive planning programmes;
b) to set up within its territories, where such services do not exist, one or more services for the protection, conservation and presentation of the cultural and natural heritage with an appropriate staff and possessing the means to discharge their functions;
c) to develop scientific and technical studies and research and to work out such operating methods as will make the State capable of counteracting the dangers that threaten its cultural or natural heritage;
d) to take the appropriate legal, scientific, technical, administrative and financial measures necessary for the identification, protection, conservation, presentation and rehabilitation of this heritage; and

e) to foster the establishment or development of national or regional centres for training in the protection, conservation and presentation of the cultural and natural heritage and to encourage scientific research in this field.

Article 6

1. Whilst fully respecting the sovereignty of the States on whose territory the cultural and natural heritage mentioned in Articles 1 and 2 is situated, and without prejudice to property right provided by national legislation, the States Parties to this Convention recognize that such heritage constitutes a world heritage for whose protection it is the duty of the international community as a whole to co-operate.
2. The States Parties undertake, in accordance with the provisions of this Convention, to give their help in the identification, protection, conservation and presentation of the cultural and natural heritage referred to in paragraphs 2 and 4 of Article 11 if the States on whose territory it is situated so request.
3. Each State Party to this Convention undertakes not to take any deliberate measures which might damage directly or indirectly the cultural and natural heritage referred to in Articles 1 and 2 situated on the territory of other States Parties to this Convention.

Article 7

For the purpose of this Convention, international protection of the world cultural and natural heritage shall be understood to mean the establishment of a system of international co-operation and assistance designed to support States Parties to the Convention in their efforts to conserve and identify that heritage.

III INTERGOVERNMENTAL COMMITTEE FOR THE PROTECTION OF THE WORLD CULTURAL AND NATURAL HERITAGE

Article 8

1. An Intergovernmental Committee for the Protection of the Cultural and Natural Heritage of Outstanding Universal Value, called "the World Heritage Committee", is hereby established within the United Nations Educational, Scientific and Cultural Organization. It shall be composed of 15 States Parties to the Convention, elected by States Parties to the Convention meeting in general assembly during the ordinary session of the General Conference of the United Nations Educational, Scientific and Cultural Organization. The number of States members of the Committee shall be increased to 21 as from the date of the ordinary session of the

General Conference following the entry into force of this Convention for at least 40 States.

2. Election of members of the Committee shall ensure an equitable representation of the different regions and cultures of the world.
3. A representative of the International Centre for the Study of the Preservation and Restoration of Cultural Property (Rome Centre), a representative of the International Council of Monuments and Sites (ICOMOS) and a representative of the International Union for Conservation of Nature and Natural Resources (IUCN), to whom may be added, at the request of States Parties to the Convention meeting in general assembly during the ordinary sessions of the General Conference of the United Nations Educational, Scientific and Cultural Organization, representatives of other intergovernmental or non-governmental organizations, with similar objectives, may attend the meetings of the Committee in an advisory capacity.

Article 9

1. The term of office of States members of the World Heritage Committee shall extend from the end of the ordinary session of the General Conference during which they are elected until the end of its third subsequent ordinary session.
2. The term of office of one-third of the members designated at the time of the first election shall, however, cease at the end of the first ordinary session of the General Conference following that at which they were elected; and the term of office of a further third of the members designated at the same time shall cease at the end of the second ordinary session of the General Conference following that at which they were elected. The names of these members shall be chosen by lot by the President of the General Conference of the United Nations Educational, Scientific and Cultural Organization after the first election.
3. States members of the Committee shall choose as their representatives persons qualified in the field of the cultural or natural heritage.

Article 10

1. The World Heritage Committee shall adopt its Rules of Procedure.
2. The Committee may at any time invite public or private organizations or individuals to participate in its meetings for consultation on particular problems.
3. The Committee may create such consultative bodies as it deems necessary for the performance of its functions.

Article 11

1. Every State Party to this Convention shall, in so far as possible, submit to the World Heritage Committee an inventory of property forming part of the cultural and natural heritage, situated in its territory and suitable for inclusion in the list provided for in paragraph 2 of this Article. This inventory, which shall not be considered exhaustive, shall include documentation about the location of the property in question and its significance.
2. On the basis of the inventories submitted by States in accordance with paragraph 1, the Committee shall establish, keep up to date and publish, under the title of "World Heritage List," a list of properties forming part of the cultural heritage and natural heritage, as defined in Articles 1 and 2 of this Convention, which it considers as having outstanding universal value in terms of such criteria as it shall have established. An updated list shall be distributed at least every two years.
3. The inclusion of a property in the World Heritage List requires the consent of the State concerned. The inclusion of a property situated in a territory, sovereignty or jurisdiction over which is claimed by more than one State shall in no way prejudice the rights of the parties to the dispute.
4. The Committee shall establish, keep up to date and publish, whenever circumstances shall so require, under the title of "list of World Heritage in Danger", a list of the property appearing in the World Heritage List for the conservation of which major operations are necessary and for which assistance has been requested under this Convention. This list shall contain an estimate of the cost of such operations. The list may include only such property forming part of the cultural and natural heritage as is threatened by serious and specific dangers, such as the threat of disappearance caused by accelerated deterioration, large-scale public or private projects or rapid urban or tourist development projects; destruction caused by changes in the use or ownership of the land; major alterations due to unknown causes; abandonment for any reason whatsoever; the outbreak or the threat of an armed conflict; calamities and cataclysms; serious fires, earthquakes, landslides; volcanic eruptions; changes in water level, floods and tidal waves. The Committee may at any time, in case of urgent need, make a new entry in the List of World Heritage in Danger and publicize such entry immediately.
5. The Committee shall define the criteria on the basis of which a property belonging to the cultural or natural heritage may be included in either of the lists mentioned in paragraphs 2 and 4 of this article.
6. Before refusing a request for inclusion in one of the two lists mentioned in paragraphs 2 and 4 of this article, the Committee shall consult the State Party in whose territory the cultural or natural property in question is situated.

7. The Committee shall, with the agreement of the States concerned, co-ordinate and encourage the studies and research needed for the drawing up of the lists referred to in paragraphs 2 and 4 of this article.

Article 12

The fact that a property belonging to the cultural or natural heritage has not been included in either of the two lists mentioned in paragraphs 2 and 4 of Article 11 shall in no way be construed to mean that it does not have an outstanding universal value for purposes other than those resulting from inclusion in these lists.

Article 13

1. The World Heritage Committee shall receive and study requests for international assistance formulated by States Parties to this Convention with respect to property forming part of the cultural or natural heritage, situated in their territories, and included or potentially suitable for inclusion in the lists mentioned referred to in paragraphs 2 and 4 of Article 11. The purpose of such requests may be to secure the protection, conservation, presentation or rehabilitation of such property.
2. Requests for international assistance under paragraph 1 of this article may also be concerned with identification of cultural or natural property defined in Articles 1 and 2, when preliminary investigations have shown that further inquiries would be justified.
3. The Committee shall decide on the action to be taken with regard to these requests, determine where appropriate, the nature and extent of its assistance, and authorize the conclusion, on its behalf, of the necessary arrangements with the government concerned.
4. The Committee shall determine an order of priorities for its operations. It shall in so doing bear in mind the respective importance for the world cultural and natural heritage of the property requiring protection, the need to give international assistance to the property most representative of a natural environment or of the genius and the history of the peoples of the world, the urgency of the work to be done, the resources available to the States on whose territory the threatened property is situated and in particular the extent to which they are able to safeguard such property by their own means.
5. The Committee shall draw up, keep up to date and publicize a list of property for which international assistance has been granted.
6. The Committee shall decide on the use of the resources of the Fund established under Article 15 of this Convention. It shall seek ways of increasing these resources and shall take all useful steps to this end.

7. The Committee shall co-operate with international and national governmental and non-governmental organizations having objectives similar to those of this Convention. For the implementation of its programmes and projects, the Committee may call on such organizations, particularly the International Centre for the Study of the Preservation and Restoration of Cultural Property (the Rome Centre), the International Council of Monuments and Sites (ICOMOS) and the International Union for Conservation of Nature and Natural Resources (IUCN), as well as on public and private bodies and individuals.
8. Decisions of the Committee shall be taken by a majority of two-thirds of its members present and voting. A majority of the members of the Committee shall constitute a quorum.

Article 14

1. The World Heritage Committee shall be assisted by a Secretariat appointed by the Director-General of the United Nations Educational, Scientific and Cultural Organization.
2. The Director-General of the United Nations Educational, Scientific and Cultural Organization, utilizing to the fullest extent possible the services of the International Centre for the Study of the Preservation and the Restoration of Cultural Property (the Rome Centre), the International Council of Monuments and Sites (ICOMOS) and the International Union for Conservation of Nature and Natural Resources (IUCN) in their respective areas of competence and capability, shall prepare the Committee's documentation and the agenda of its meetings and shall have the responsibility for the implementation of its decisions.

IV FUND FOR THE PROTECTION OF THE WORLD CULTURAL AND NATURAL HERITAGE

Article 15

1. A Fund for the Protection of the World Cultural and Natural Heritage of Outstanding Universal Value, called "the World Heritage Fund", is hereby established.
2. The Fund shall constitute a trust fund, in conformity with the provisions of the Financial Regulations of the United Nations Educational, Scientific and Cultural Organization.
3. The resources of the Fund shall consist of:
    a) compulsory and voluntary contributions made by States Parties to this Convention,
    b) Contributions, gifts or bequests which may be made by:
        i. other States;

   ii. the United Nations Educational, Scientific and Cultural Organization, other organizations of the United Nations system, particularly the United Nations Development Programme or other intergovernmental organizations;
   iii. public or private bodies or individuals;

c) any interest due on the resources of the Fund;
d) funds raised by collections and receipts from events organized for the benefit of the fund; and
e) All other resources authorized by the Fund's regulations, as drawn up by the World Heritage Committee.

4. Contributions to the Fund and other forms of assistance made available to the Committee may be used only for such purposes as the Committee shall define. The Committee may accept contributions to be used only for a certain programme or project, provided that the Committee shall have decided on the implementation of such programme or project. No political conditions may be attached to contributions made to the Fund.

Article 16

1. Without prejudice to any supplementary voluntary contribution, the States Parties to this Convention undertake to pay regularly, every two years, to the World Heritage Fund, contributions, the amount of which, in the form of a uniform percentage applicable to all States, shall be determined by the General Assembly of States Parties to the Convention, meeting during the sessions of the General Conference of the United Nations Educational, Scientific and Cultural Organization. This decision of the General Assembly requires the majority of the States Parties present and voting, which have not made the declaration referred to in paragraph 2 of this Article. In no case shall the compulsory contribution of States Parties to the Convention exceed 1% of the contribution to the regular budget of the United Nations Educational, Scientific and Cultural Organization.
2. However, each State referred to in Article 31 or in Article 32 of this Convention may declare, at the time of the deposit of its instrument of ratification, acceptance or accession, that it shall not be bound by the provisions of paragraph 1 of this Article.
3. A State Party to the Convention which has made the declaration referred to in paragraph 2 of this Article may at any time withdraw the said declaration by notifying the Director-General of the United Nations Educational, Scientific and Cultural Organization. However, the withdrawal of the declaration shall not take effect in regard to the compulsory contribution due by the State until the date of the subsequent General Assembly of States parties to the Convention.
4. In order that the Committee may be able to plan its operations effectively, the contributions of States Parties to this Convention which have made

the declaration referred to in paragraph 2 of this Article, shall be paid on a regular basis, at least every two years, and should not be less than the contributions which they should have paid if they had been bound by the provisions of paragraph 1 of this Article.

5. Any State Party to the Convention which is in arrears with the payment of its compulsory or voluntary contribution for the current year and the calendar year immediately preceding it shall not be eligible as a Member of the World Heritage Committee, although this provision shall not apply to the first election.

The terms of office of any such State which is already a member of the Committee shall terminate at the time of the elections provided for in Article 8, paragraph 1 of this Convention.

Article 17
The States Parties to this Convention shall consider or encourage the establishment of national public and private foundations or associations whose purpose is to invite donations for the protection of the cultural and natural heritage as defined in Articles 1 and 2 of this Convention.

Article 18
The States Parties to this Convention shall give their assistance to international fund-raising campaigns organized for the World Heritage Fund under the auspices of the United Nations Educational, Scientific and Cultural Organization. They shall facilitate collections made by the bodies mentioned in paragraph 3 of Article 15 for this purpose.

V CONDITIONS AND ARRANGEMENTS FOR INTERNATIONAL ASSISTANCE

Article 19
Any State Party to this Convention may request international assistance for property forming part of the cultural or natural heritage of outstanding universal value situated within its territory. It shall submit with its request such information and documentation provided for in Article 21 as it has in its possession and as will enable the Committee to come to a decision.

Article 20
Subject to the provisions of paragraph 2 of Article 13, sub-paragraph (c) of Article 22 and Article 23, international assistance provided for by this Convention may be granted only to property forming part of the cultural and natural heritage which the World Heritage Committee has decided, or may

decide, to enter in one of the lists mentioned in paragraphs 2 and 4 of Article 11.

Article 21

1. The World Heritage Committee shall define the procedure by which requests to it for international assistance shall be considered and shall specify the content of the request, which should define the operation contemplated, the work that is necessary, the expected cost thereof, the degree of urgency and the reasons why the resources of the State requesting assistance do not allow it to meet all the expenses. Such requests must be supported by experts' reports whenever possible.
2. Requests based upon disasters or natural calamities should, by reasons of the urgent work which they may involve, be given immediate, priority consideration by the Committee, which should have a reserve fund at its disposal against such contingencies.
3. Before coming to a decision, the Committee shall carry out such studies and consultations as it deems necessary.

Article 22

Assistance granted by the World Heritage Fund may take the following forms:

a) studies concerning the artistic, scientific and technical problems raised by the protection, conservation, presentation and rehabilitation of the cultural and natural heritage, as defined in paragraphs 2 and 4 of Article 11 of this Convention;
b) provisions of experts, technicians and skilled labour to ensure that the approved work is correctly carried out;
c) training of staff and specialists at all levels in the field of identification, protection, conservation, presentation and rehabilitation of the cultural and natural heritage;
d) supply of equipment which the State concerned does not possess or is not in a position to acquire;
e) low-interest or interest-free loans which might be repayable on a long-term basis;
f) the granting, in exceptional cases and for special reasons, of non-repayable subsidies.

Article 23

The World Heritage Committee may also provide international assistance to national or regional centres for the training of staff and specialists at all levels in the field of identification, protection, conservation, presentation and rehabilitation of the cultural and natural heritage.

Article 24

International assistance on a large scale shall be preceded by detailed scientific, economic and technical studies. These studies shall draw upon the most advanced techniques for the protection, conservation, presentation and rehabilitation of the natural and cultural heritage and shall be consistent with the objectives of this Convention. The studies shall also seek means of making rational use of the resources available in the State concerned.

Article 25

As a general rule, only part of the cost of work necessary shall be borne by the international community. The contribution of the State benefiting from international assistance shall constitute a substantial share of the resources devoted to each programme or project, unless its resources do not permit this.

Article 26

The World Heritage Committee and the recipient State shall define in the agreement they conclude the conditions in which a programme or project for which international assistance under the terms of this Convention is provided, shall be carried out. It shall be the responsibility of the State receiving such international assistance to continue to protect, conserve and present the property so safeguarded, in observance of the conditions laid down by the agreement.

VI EDUCATIONAL PROGRAMMES

Article 27

1. The States Parties to this Convention shall endeavor by all appropriate means, and in particular by educational and information programmes, to strengthen appreciation and respect by their peoples of the cultural and natural heritage defined in Articles 1 and 2 of the Convention.
2. They shall undertake to keep the public broadly informed of the dangers threatening this heritage and of the activities carried on in pursuance of this Convention.

Article 28

States Parties to this Convention which receive international assistance under the Convention shall take appropriate measures to make known the importance of the property for which assistance has been received and the role played by such assistance.

VII REPORTS

Article 29

1. The States Parties to this Convention shall, in the reports which they submit to the General Conference of the United Nations Educational, Scientific and Cultural Organization on dates and in a manner to be determined by it, give information on the legislative and administrative provisions which they have adopted and other action which they have taken for the application of this Convention, together with details of the experience acquired in this field.
2. These reports shall be brought to the attention of the World Heritage Committee.
3. The Committee shall submit a report on its activities at each of the ordinary sessions of the General Conference of the United Nations Educational, Scientific and Cultural Organization.

VIII FINAL CLAUSES

Article 30

This Convention is drawn up in Arabic, English, French, Russian and Spanish, the five texts being equally authoritative.

Article 31

1. This Convention shall be subject to ratification or acceptance by States members of the United Nations Educational, Scientific and Cultural Organization in accordance with their respective constitutional procedures.
2. The instruments of ratification or acceptance shall be deposited with the Director-General of the United Nations Educational, Scientific and Cultural Organization.

Article 32

1. This Convention shall be open to accession by all States not members of the United Nations Educational, Scientific and Cultural Organization which are invited by the General Conference of the Organization to accede to it.
2. Accession shall be effected by the deposit of an instrument of accession with the Director-General of the United Nations Educational, Scientific and Cultural Organization.

Article 33
This Convention shall enter into force three months after the date of the deposit of the twentieth instrument of ratification, acceptance or accession, but only with respect to those States which have deposited their respective instruments of ratification, acceptance or accession on or before that date. It shall enter into force with respect to any other State three months after the deposit of its instrument of ratification, acceptance or accession.

Article 34
The following provisions shall apply to those States Parties to this Convention which have a federal or non-unitary constitutional system:

a) with regard to the provisions of this Convention, the implementation of which comes under the legal jurisdiction of the federal or central legislative power, the obligations of the federal or central government shall be the same as for those States parties which are not federal States;
b) with regard to the provisions of this Convention, the implementation of which comes under the legal jurisdiction of individual constituent States, countries, provinces or cantons that are not obliged by the constitutional system of the federation to take legislative measures, the federal government shall inform the competent authorities of such States, countries, provinces or cantons of the said provisions, with its recommendation for their adoption.

Article 35
1. Each State Party to this Convention may denounce the Convention.
2. The denunciation shall be notified by an instrument in writing, deposited with the Director-General of the United Nations Educational, Scientific and Cultural Organization.
3. The denunciation shall take effect twelve months after the receipt of the instrument of denunciation. It shall not affect the financial obligations of the denouncing State until the date on which the withdrawal takes effect.

Article 36
The Director-General of the United Nations Educational, Scientific and Cultural Organization shall inform the States members of the Organization, the States not members of the Organization which are referred to in Article 32, as well as the United Nations, of the deposit of all the instruments of ratification, acceptance, or accession provided for in Articles 31 and 32, and of the denunciations provided for in Article 35.

Article 37

1. This Convention may be revised by the General Conference of the United Nations Educational, Scientific and Cultural Organization. Any such revision shall, however, bind only the States which shall become Parties to the revising convention.
2. If the General Conference should adopt a new convention revising this Convention in whole or in part, then, unless the new convention otherwise provides, this Convention shall cease to be open to ratification, acceptance or accession, as from the date on which the new revising convention enters into force.

Article 38

In conformity with Article 102 of the Charter of the United Nations, this Convention shall be registered with the Secretariat of the United Nations at the request of the Director-General of the United Nations Educational, Scientific and Cultural Organization.

Done in Paris, this twenty-third day of November 1972, in two authentic copies bearing the signature of the President of the seventeenth session of the General Conference and of the Director-General of the United Nations Educational, Scientific and Cultural Organization, which shall be deposited in the archives of the United Nations Educational, Scientific and Cultural Organization, and certified true copies of which shall be delivered to all the States referred to in Articles 31 and 32 as well as to the United Nations.

Appendix II

Secretary of the Interior's Standards for the Treatment of Historic Properties

STANDARDS FOR PRESERVATION

1. A property will be used as it was historically, or be given a new use that maximizes the retention of distinctive materials, features, spaces, and spatial relationships. Where a treatment and use have not been identified, a property will be protected and, if necessary, stabilized until additional work may be undertaken.
2. The historic character of a property will be retained and preserved. The replacement of intact or repairable historic materials or alteration of features, spaces, and spatial relationships that characterize a property will be avoided.
3. Each property will be recognized as a physical record of its time, place, and use. Work needed to stabilize, consolidate, and conserve existing historic materials and features will be physically and visually compatible, identifiable upon close inspection, and properly documented for future research.
4. Changes to a property that have acquired historic significance in their own right will be retained and preserved.
5. Distinctive materials, features, finishes, and construction techniques or examples of craftsmanship that characterize a property will be preserved.
6. The existing condition of historic features will be evaluated to determine the appropriate level of intervention needed. Where the severity of deterioration requires repair or limited replacement of a distinctive feature, the new material will match the old in composition, design, color, and texture.

7. Chemical or physical treatments, if appropriate, will be undertaken using the gentlest means possible. Treatments that cause damage to historic materials will not be used.
8. Archaeological resources will be protected and preserved in place. If such resources must be disturbed, mitigation measures will be undertaken.

STANDARDS FOR REHABILITATION

1. A property will be used as it was historically or be given a new use that requires minimal change to its distinctive materials, features, spaces, and spatial relationships.
2. The historic character of a property will be retained and preserved. The removal of distinctive materials or alteration of features, spaces, and spatial relationships that characterize a property will be avoided.
3. Each property will be recognized as a physical record of its time, place, and use. Changes that create a false sense of historical development, such as adding conjectural features or elements from other historic properties, will not be undertaken.
4. Changes to a property that have acquired historic significance in their own right will be retained and preserved.
5. Distinctive materials, features, finishes, and construction techniques or examples of craftsmanship that characterize a property will be preserved.
6. Deteriorated historic features will be repaired rather than replaced. Where the severity of deterioration requires replacement of a distinctive feature, the new feature will match the old in design, color, texture, and, where possible, materials. Replacement of missing features will be substantiated by documentary and physical evidence.
7. Chemical or physical treatments, if appropriate, will be undertaken using the gentlest means possible. Treatments that cause damage to historic materials will not be used.
8. Archaeological resources will be protected and preserved in place. If such resources must be disturbed, mitigation measures will be undertaken.
9. New additions, exterior alterations, or related new construction will not destroy historic materials, features, and spatial relationships that characterize the property. The new work shall be differentiated from the old and will be compatible with the historic materials, features, size, scale and proportion, and massing to protect the integrity of the property and its environment.

10. New additions and adjacent or related new construction will be undertaken in such a manner that, if removed in the future, the essential form and integrity of the historic property and its environment would be unimpaired.

STANDARDS FOR RESTAURATION

1. A property will be used as it was historically or be given a new use which reflects the property's restoration period.
2. Materials and features from the restoration period will be retained and preserved. The removal of materials or alteration of features, spaces, and spatial relationships that characterize the period will not be undertaken.
3. Each property will be recognized as a physical record of its time, place, and use. Work needed to stabilize, consolidate and conserve materials and features from the restoration period will be physically and visually compatible, identifiable upon close inspection, and properly documented for future research.
4. Materials, features, spaces, and finishes that characterize other historical periods will be documented prior to their alteration or removal.
5. Distinctive materials, features, finishes, and construction techniques or examples of craftsmanship that characterize the restoration period will be preserved.
6. Deteriorated features from the restoration period will be repaired rather than replaced. Where the severity of deterioration requires replacement of a distinctive feature, the new feature will match the old in design, colour, texture, and, where possible, materials.
7. Replacement of missing features from the restoration period will be substantiated by documentary and physical evidence. A false sense of history will not be created by adding conjectural features, features from other properties, or by combining features that never existed together historically.
8. Chemical or physical treatments, if appropriate, will be undertaken using the gentlest means possible. Treatments that cause damage to historic materials will not be used.
9. Archaeological resources affected by a project will be protected and preserved in place. If such resources must be disturbed, mitigation measures will be undertaken.
10. Designs that were never executed historically will not be constructed.

STANDARDS FOR RECONSTRUCTION

1. Reconstruction will be used to depict vanished or non-surviving portions of a property when documentary and physical evidence is available to permit accurate reconstruction with minimal conjecture, and such reconstruction is essential to the public understanding of the property.
2. Reconstruction of a landscape, building, structure, or object in its historic location will be preceded by a thorough archaeological investigation to identify and evaluate those features and artefacts which are essential to an accurate reconstruction. If such resources must be disturbed, mitigation measures will be undertaken.
3. Reconstruction will include measures to preserve any remaining historic materials, features, and spatial relationships.
4. Reconstruction will be based on the accurate duplication of historic features and elements substantiated by documentary or physical evidence rather than on conjectural designs or the availability of different features from other historic properties. A reconstructed property will re-create the appearance of the non-surviving historic property in materials, design, colour, and texture.
5. A reconstruction will be clearly identified as a contemporary re-creation.
6. Designs that were never executed historically will not be constructed.

Bibliography

ABAD LICERAS, J.M. "El principio jurisprudencial del derecho social a la cultura: análisis y manifestaciones" in Patrimonio Cultural y Derecho **7**, 2002.

ABAD LICERAS, J.M. "El décimo aniversario de la ley 16 / 1985 del Patrimonio Histórico Español", Boletín del Instituto Andaluz del Patrimonio Histórico **11**, 1995, pp. 22–25.

ABAD LICERAS, J.M. "¿Está el Patrimonio Histórico suficientemente protegido en el código penal?", R&R: Restauración y Rehabilitación **10**, 1997, pp. 54–57.

ABAD LICERAS, J.M. "Las medidas de fomento del patrimonio histórico en la legislación española y europea", R&R: Restauración y Rehabilitación **3**, 1997, pp. 58–60.

ABAD LICERAS, J.M. Fundaciones, asociaciones y patrimonio histórico, R&R: Restauración y rehabilitación **24**, 1999, pp. 58–59.

ABAD LICERAS, J.M. La situación de ruina y demolición de inmuebles del Patrimonio Histórico, Montecorvo, Madrid, 2000.

AGUDO TORRICO, J. "Patrimonio etnológico. Problemática en torno a su definición y objetivos", Boletín del Instituto Andaluz del Patrimonio Histórico **18**, 1997, pp. 97–108.

AGUILAR CIVERA, I. "El patrimonio arquitectónico industrial. Una reflexión a partir de la actividad Valenciana", Boletín del Instituto Andaluz del Patrimonio Histórico **21**, 1997, pp. 118–124.

AGUILAR CIVERA, I. "La investigación del Patrimonio Histórico industrial del Puerto de Sagunto" En Reconversión y Revolución, Industrialización y Patrimonio, Universidad de Valencia, Valencia, 1999, p. 37.

AHERN K. Cultural Landscapes Bibliography: An Annotated Bibliography on Resources in the National Park System, Washington DC: USDI, NPS, Cultural Resources, Park Historic Architecture Division, 1992.

ALANEN, A.R. & MELNICK, R.Z. (eds) Preserving Cultural Landscapes in America. Baltimore, Md. Johns Hopkins University Press, 2000.

ALANEN, A.R. *et al.* Preserving Cultural Landscapes in America. Baltimore: Johns Hopkins University Press, 2000.

ALBANO, A. & PAGANO. E. (eds) Legislazione internazionale e comunitaria dei beni culturali Ed. Giuridiche Simone, Gruppo Editoriale Esselibri, 2005, pp. 269.

ALEGRE ÁVILA, J.M. Evolución y régimen jurídico del Patrimonio Histórico, Ministerio de Cultura, Madrid, 1994.

ALFRY, J. & PUTNAM, T. The Industrial Heritage, Managing Resources and Uses, Routledge, 1992.

ALIBRANDI, TOMMASO, FERRI, PIER GIORGIO, IL diritto dei beni culturali: la protezione del patrimonio storico–artístico, La Nuova Italia Scientifica, Roma, 1988.

ALLEN, T. Regional Heritage Areas: Approaches to Sustainable Development. National Trust for Historic Preservation, 1994.

ALONSO DE ARMIÑO, L. "La excepcionalidad de la vivienda social y obrera en el Puerto de Sagunto" En Reconversión y Revolución, Industrialización y Patrimonio, Universidad de Valencia, Valencia, 1999.

ALONSO IBAÑEZ, M.R. "La identificación de los espacios culturales en el ordenamiento español y la necesidad de integrar su tratamiento en el marco de la ordenación territorial", Patrimonio Cultural y Derecho **1**, 1997, pp. 101–119.

ALONSO IBAÑEZ, M.R. "Patrimonio Industrial. Notas a su insatisfactoria protección jurídica" en Revista de Patrimonio Cultural y Derecho **3**, 1999, pp. 257–275.

ALONSO IBAÑEZ, M.R. "Un nuevo supuesto de incumplimiento del Derecho Comunitario por el Estado español: el régimen legal de visita pública de los bienes de interés cultural", Patrimonio Cultural y Derecho **1**, 1997, pp. 209–211.

ALONSO IBAÑEZ, M.R. El Patrimonio Histórico: destino público y valor cultural, Civitas, Madrid, 1992.

ALTED VIGIL, A. Política del nuevo Estado sobre el patrimonio cultural y la educación durante la Guerra Civil Española, Centro Nacional de Información Artística, Arqueológica y Etnológica, Madrid, 1984.

ÁLVAREZ ÁLVAREZ, J.L. "El patrimonio cultural. De dónde venimos, dónde estamos y a dónde vamos", Patrimonio Cultural y Derecho **1**, 1997, pp. 15–31.

ÁLVAREZ ÁLVAREZ, J.L. "La defensa del patrimonio cultural y la Real Academia de Bellas Artes de San Fernando", Patrimonio Cultural y Derecho **2**, 1998, pp. 165–170.

ÁLVAREZ ÁLVAREZ, J.L. Estudios sobre el Patrimonio Histórico Español, Civitas, Madrid, 1989.

ÁLVAREZ LOPERA, J. "La organización de la defensa de bienes culturales en Cataluña. El periodo revolucionario (Julio 1936 – Junio 1937)", Cuadernos de Arte de la Universidad de Granada **16**, 1984, pp. 533–592.

ÁLVAREZ MARTÍNEZ, M.S. Santa Cristina de Lena. [Folleto], Servicio de Publicaciones del Principado de Asturias, Oviedo, 1988.

AMADOR DE LOS RIOS, R. Catálogo de los monumentos históricos y artísticos de la provincia de Huelva, Carrasco Terriza, Manuel Jesús, Diputación Provincial de Huelva, 1998.

ANDERSON, M. "In search of woman's public history: heritage and gender" in Public History Review **3**.

ASHWORTH & HOWERY European Heritage Planning and Management, Intellect Books, 1999.

BARRERO RODRÍGUEZ, M.C. "La organización administrativa de las bellas artes: unas reflexiones de futuro", Patrimonio Cultural y Derecho **1**, 1997, pp. 75-99.

BARRERO RODRÍGUEZ, M.C. "Normas materiales en ordenación urbanística. A propósito de las remodelaciones urbanas en conjuntos históricos" en El Derecho Administrativo en el umbral del siglo XXI. Libro Homenaje al Profesor Dr. D. Ramón Martín Mateo, Tirant lo Blanch, Valencia, 2000, pp. 3333–3363.

BARRERO RODRÍGUEZ, M.C. La ordenación jurídica del Patrimonio Histórico Madrid, Civitas, 1990.

BARRETT, T. & NAGEL, S. Model Conservation Easement and Historic Preservation Easement, Revised Easement and Commentary, 1996.

BENÍTEZ DE LUGO E.L. "Servicios Territoriales de Patrimonio Histórico", Boletín del Instituto Andaluz del Patrimonio Histórico **14**, 1996, pp. 20–23.

BENITEZ DE LUGO E. & GUILLEN, F. El Patrimonio Cultural Español, Ed. Comares, Granada, 1995.

BENSON J.F. & ROE M.H. Landscape and Sustainability, London and New York, 2000.

BERNHARDT, R. & BURKHART, A. Real Property in a Nutshell, 2000.

BONIFACE, P. & FOWLER, P.J. Heritage and Tourism in the 'global village', London: Routledge, 1993.

BRADY, J. "Land is itself a sacred, living being. Native American sacred site protection on federal public lands amidst the shadows of Bear Lodge" in 24 Am. Indian L. Rev. 153.

BRAY, P. "Rethinking Urban Parks." Places **16(**2), Spring 2004.

BRU I VIDAL, S. Sagunto–El Puig, Conselleria de Cultura, Educació i Ciència, Valencia, 1984.

BRUCE, J. & ELY, J. The Law of Easements and Licenses in Land, 1988.

BYRNE, C.L. & MINCK, M. "Understanding the Evolution of Conservation Easement Appraisal through Case Law," 411 The Appraisal Journal, October 2000.

CAMPILLO GARRIGÓS, R. La gestión y el gestor del Patrimonio Cultural, KR, Murcia, 1998.

CANO MURCIA, M. "El interdicto de obra ruinosa y el procedimiento de declaración de ruina", La Ley **2714** de 10 de Mayo, 1991.

CANTUCCI, M. La tutela giuridica dell cose d'interesse artístico e storico, Cedam, Padova, 1953.

CARTIER, C. Seductions of Place: Geographical Perspectives on Globalization and Touristed Landscapes Charlottesville: University of Virginia Press, 2005.

CASANELLES I RAHOLA, E. "Por un nuevo marco legal del patrimonio: el caso del patrimonio industrial", Patrimonio Cultural y Derecho **1**, 1997, pp. 193–198.

CASANELLES I RAHOLA, E. "Patrimonio industrial y museológico en Cataluña", Boletín del Instituto Andaluz del Patrimonio Histórico **21**, 1997, pp. 125–129.

CASTELLANO GAMEZ, M. & SÁNCHEZ MARTÍNEZ, J.A. "Apuntes para la gestion del patrimonio historico desde una perspectiva municipalista", Boletín del Instituto Andaluz del Patrimonio Histórico **17**, 1996, pp. 90–95.

CASTILLO RUÍZ, J. "¿Hacia una nueva definición del patrimonio histórico? Reflexiones sobre el documento "bases para una carta sobre patrimonio y desarrollo en Andalucía", Boletín del Instituto Andaluz del Patrimonio Histórico **16**, 1996, pp. 101–106.

CASTILLO RUÍZ, J. "La Ley de Patrimonio Histórico de Andalucía como instrumento de tutela", Boletín del Instituto Andaluz del Patrimonio Histórico **13**, 1995, pp. 30–35.

CASTILLO RUÍZ, J. "El entorno de los bienes inmuebles de interés cultural: su regulación en la Ley de Patrimonio Histórico Español", Cuadernos de Arte de la Universidad de Granada **23**, 1992, pp. 615–624.

CHAPMAN, T. in "To save and save not: the historic preservation implications of the property rights movement" in 77 B.U.L. Rev. 111 Feb. 1997.

CHINCHILLA MARÍN, C. & PRIETO DE TORMO, J. "Las nuevas fronteras del concepto de patrimonio cultural: el toro de Osborne". Patrimonio Cultural y Derecho **1**, 1997, pp. 279–298.

CHUMLEY, C. & UTT, R. National Heritage Areas: Costly Economic Development Schemes that Threaten Property Rights, in Backgrounder No. 2080 October 23, National Heritage Foundation, 2007.

CLARK, K. Informed Conservation: understanding historic buildings and their landscapes for conservation, London: English Heritage, 2000

COCCOSSIS, H. & PARPAIRIS, A. "Assessing the interaction between heritage, environment and tourism", Mykonos, Sustainable Tourism Development, pp. 107-125, Hants (England), Avebury, 1995.

COLLEY, S. Uncovering Australia, Archaeology, Indigenous People and the Public, Allen & Unwin, 2002.

CREMIN, D.H. "From Heartland to Nation: The National Heritage Area Movement at Twenty." History News **59**(2), 2004.

CREW, S. & SIMS, J. Locating Authenticity: fragments of a dialogue, in "Exhibiting Cultures: the poetics and politics of a museum display", Washington DC, 1991.

DEPARTMENT FOR COMMUNITIES AND LOCAL GOVERNMENT "Revisions to principles of selection for listing buildings" March 2007.

DEPARTMENT FOR COMMUNITIES AND LOCAL GOVERNMENT Best Practice Guidance on Listed Buildings Prosecutions, London, December 2006.

DEPARTMENT FOR COMMUNITIES AND LOCAL GOVERNMENT Circular 01/07 Revisions to Principles of Selection for Listed Buildings, March 2007.

DIEHL, J. & BARRETT, T. The Conservation Easement Handbook: Managing land Conservation and Historic Preservation Easement Programs. San Francisco: Trust for Public Land, 1988.

DUERKSEN, C.J. A Handbook on Historic Preservation Law. Washington, D.C., Conservation Foundation, 1983.

DUPONT, D. "Colloque International sur l'Interpretation du Patrimoine (1°. 1996. Auvers–SurOise)", Premier colloque international sur l'Interprétation du patrimoine: Château d'Auvers (Auvers–sur–Oise) 3 et 4 Décembre 1996, Editions Touristiques Européennes, París, 1997.

DUPRÉ I RAVENTOS, X. "El consell d'europa recomenda la proteccio del patrimoni arqueologic en el context de les operacion urbanistiques", Espais: Revista del Departament de Política Territorial i Obres Públiques **19**, 1989, pp. 7–13.

ELIA, RICARDO J. "US Cultural Resource Management and the ICAHM Charter," Antiquity **67(**255), 1993, pp. 426–438.

ENGLISH HERITAGE, HERITAGE PROTECTION DEPARTMENT Industrial Buildings Selection Guide, March 2007.

ESCALERA REYES, J. "La fiesta como patrimonio", Boletín del Instituto Andaluz del Patrimonio Histórico **21**, 1997, pp. 53–58.

EVANS, M. Principles of Environmental Heritage and Law, Prospect, Sydney 2000.

FERNÁNDEZ–BACA CASARES, R. "Patrimonio histórico, cohesión social e innovación", Boletín del Instituto Andaluz del Patrimonio Histórico **27**, 1999, pp. 118–123.

FERNÁNDEZ FERNÁNDEZ, J. "La conservación de los bienes culturales y la ley 16/1985, de 25 de junio, de Patrimonio Histórico Español", VI Congreso de Conservación de Bienes Culturales: Tarragona, 29 de mayo al 1 de junio 1986, Tarragona, 1986, pp. 31–41.

FERNÁNDEZ NARANJO, J.A. "La conservación de los bienes inmuebles de interés cultural pertenecientes al patrimonio histórico en Andalucía", Boletín del Instituto Andaluz del Patrimonio Histórico **25**, 1998, pp. 97–104.

FERNÁNDEZ RODRÍGUEZ, T.R. "Legislación española sobre protección del patrimonio artístico. Balance de cara a su reforma", Revista de Derecho Urbanístico **60**, 1978.

FITZ GIBBON, K. (ed.) Who owns the past?: cultural policy, cultural property, and the law, Rutgers University Press, 2005, pp. xx, 362 p.

FOWLER, P.J. 'Cultural landscapes of Britain', International Journal Heritage Studies **6**, 2000, pp. 201–12.

FRANCIONI, F. "Treinta años después ¿está la Convención sobre el Patrimonio Mundial preparada para el siglo XXI" in Patrimonio Cultural y Derecho 7, 2003 p. 12.

GARCÍA DE ENTERRIA, E. "Consideraciones sobre una nueva legislación del Patrimonio artístico, histórico y cultural" REDA **39**, 1983, pp. 578–610.

GARCÍA-ESCUDERO, P. & PENDAS GARCÍA, B. El nuevo régimen jurídico del patrimonio histórico español, Subdirección General de Documentación y Publicaciones, Madrid, 1986.

GARCÍA FERNÁNDEZ, J. "La protección jurídica del patrimonio cultural: nuevas cuestiones y nuevos sujetos a los diez años de la ley del patrimonio histórico español", Patrimonio Cultural y Derecho **1**, 1997, pp. 53–74.

GARCÍA GARCÍA, M.J. El régimen jurídico de la rehabilitación urbana, Institució Alfons el Magnanim, Diputación de Valencia, Valencia, 1999.

GARCIA MACHO, R. "De los derechos sociales a los principios rectores", Separata del libro La Declaración Universal de los Derechos Humanos en su 50 aniversario, Editorial Bosch, Barcelona, 1998.

GEMMA PRIETO, M. "Derecho Internacional y bienes culturales", Patrimonio Cultural y Derecho **2**, 1998, pp. 150–185.

GERSTENBLITH, P. "Identity and Cultural Property: The Protection of Cultural Property in the United States," Boston University Law Review **75(**3), 1995, pp. 559–688.

GIANNINI "I benni culturali" Rivista trimestrale di Diritto Público año 26, **1**, 1976.

GONZÁLES-ÚBEDA RICO Aspectos jurídicos de la protección del Patrimonio Histórico–Artístico y Cultural, Ministerio de Cultura, Madrid, 1981.

GREEN, B. & VOSS, W. Threatened Landscapes: Conserving Cultural Environments, London and New York: Spon Press 2000.

GRIMM, L. "Sacred Lands and the Establishment Clause: Indian Religious Practices on Federal Lands", 12 Nat. Resources & Environment **19,** 1997.

GUEDES, P. (ed.) The Macmillan Encyclopaedia of Architecture and Technological Change 1979.

GUERRA DE HOYOS, C. "Patrimonio cultural y reutilización", Boletín del Instituto Andaluz del Patrimonio Histórico **18**, 1997, pp. 93–96.

GUSTANSKI, J. Protecting the Land: Conservation Easements Past, Present and Future. Washington, DC: Island Press, 2000.

HARDESTY, D. & LITTLE, B. Assessing Site Significance: A Guide for Archaeologists and Historians. Heritage Resources Management Series 3, Walnut Creek, CA: AltaMira, 2000.

HART, J. "Planning for and Preserving Cultural Resources through National Heritage Areas", Cultural Resource Management **23(**7), 2000, pp. 29–32. See also: Cultural Conservation and Folkways.

HAUGHWOUT FOLSOM, R., LAKE, R. & NANDA, V. European Union Law after Maastricht: A Practical Guide for Lawyers Outside the Common Market, Kluwer Law International, 1996.

HERNÁNDEZ LEÓN, J.M. "La naturaleza de lo antiguo y el impulso restaurador", Boletín del Instituto Andaluz del Patrimonio Histórico **19**, 1997, pp. 94–99.

HERRERA VIRUMBRALES, J. "Décimo aniversario de la ley del Patrimonio Histórico Español: reflexiones en voz alta", Boletín del Instituto Andaluz del Patrimonio Histórico **12**, 1995, pp. 25–35.

HIDALGO CUÑARRO, J.M. "Coloquio Internacional sobre Conservación Preventiva de Bienes Culturales", Actas del Coloquio Internacional sobre Conservación Preventiva de Bienes Culturales, Diputación de Pontevedra, 1997.

HIRSCH, E. & O'HANLON The Anthropology of Landscape: Perspectives on Place and Space, Oxford: Clarendon Press, 1995.

HOFFMAN, B. Art and Cultural Heritage: Law, Policy and Practice, Cambridge University Press.

HUDSON, K. Industrial archaeology: an introduction, Oxford University Press, London, 1964.

HUDSON, K. Industrial Archaeology: a New Introduction, 3rd revised and reset, 1976.

HUDSON, K. The Archaeology of the Consumer Society: the Second Industrial Revolution in Britain, 1983.

INSKEEP, E. Tourism Planning: An Integrated and Sustainable Development Approach, New York, Van Nostrand Reinhold, 1991.

JONES, W.R. Dictionary of Industrial Archaeology, 1996.

KING, T. Cultural Resource Laws & Practice: An Introductory Guide. Heritage Resources Management Series 1, Walnut Creek, CA: AltaMira, 1998.

KISS, A. La notion de Patrimoine commun de l'humanité, Rec de cours de l'Academie de Droit International de La Haye II, 1982.

LA PARRA LÓPEZ, S. "Patrimoni cultural, turisme i democràcia", Espai Obert: Revista d'Assaig i Investigació (Comarques Centrals Valencianes) **5**, 1998, pp. 17–20.

LAFUENTE BATANERO, L. "El consejo del patrimonio histórico", Patrimonio Cultural y Derecho **2**, 1998, pp. 21–53.

LARKHAM, P. Conservation and the city, Routlege 1996, p. 266.

LEE, R. The Antiquities Act of 1906. National Park Service, National Technical Information Service.

LOPEZ BELTRAN DE HEREDIA, C. La Ley Valenciana de patrimonio cultural: Ley 4/1998, de 11 de junio, del Patrimonio Histórico–artístico, normas reguladora del patrimonio cultural Valenciano, Tirant lo Blanch, Valencia, 1999.

LÓPEZ BRAVO, C. "Interrelación de las categorías legales de protección del patrimonio cultural en España", Boletín del Instituto Andaluz del Patrimonio Histórico **27**, 1999, pp. 83–90.

LÓPEZ BRAVO, C. "Los bienes culturales en el Derecho estatal y autonómico de España" en Patrimonio Cultural y Derecho **3**, 1999, pp. 11–25.

LORENZO, FELICITAS, A. "Los bienes culturales y los inventarios de bienes de entidades locales en Bizkaia", Boletín del Instituto Andaluz del Patrimonio Histórico **10**, 1995, pp. 20–25.

LOUREIRO LAMAS, C. & SÁNCHEZ GARCÍA, A.M. "Patrimonio cultural, patrimonio natural", Cuadernos de Etnología y Etnografía de Navarra **56**, 1990, pp. 321–329.

LOWENTHAL, D. The past is a foreign country, Cambridge University Press, Cambridge, 1986.

LOWENTHAL, D. Introduction. In: Our Past Before Us, Why Do We Save It? London, Temple Smith, 1981.

LUIS PIÑÓN, J. "Patrimonio Histórico y proyecto urbano: El caso de Port de Sagunt", en Reconversión y Revolución, Industrialización y Patrimonio, edit. Universidad de Valencia, 1999.

MAGAN PERALES, J.M. "Algunas precisiones sobre la noción de bien cultural en el Tratado de la Comunidad Europea. La referencia al patrimonio histórico del artículo 30 del TCE (antiguo art. 36)", Patrimonio Cultural y Derecho **2**, 1998, pp. 171–182.

MANSI ANTONIO, A. Storia e legislazione dei beni culturali ambientali, Udine: Del Bianco, Roma, 1988.

MARCHENA GÓMEZ, M.J. Turismo urbano y patrimonio cultural: una perspectiva europea,Diputación Provincial de Sevilla, Sevilla, 1998.

MARTÍN-BUENO, M.A. "Patrimonio Cultural y Arqueología" AIC: Análisis e Investigaciones Culturales **24**, 1985, pp. 37–48.

MARTÍN MARTÍNEZ, J. Urbanismo y arquitectura industrial en Puerto de Sagunto: (1907–1936), Caja de Ahorros de Sagunto, Sagunto, 1990.

MARTÍN, M. Difusión del Patrimonio Industrial: la historia, Boletín del Instituto Andaluz del Patrimonio Histórico **5**, 1993, pp. 45–55.

MARTÍN MATEO, R. Tratado de Derecho Ambiental, vol. I, Trivium, Madrid, 1991. Tratado de Derecho Ambiental, vol. II, Trivium, Madrid, 1992.

MARTÍN REBOLLO, L. El comercio del arte y la Unión Europea: la normativa comunitaria sobre exportación y tráfico de bienes culturales, Madrid: Fundación Universidad Empresa, Civitas, Madrid, 1994.

MARTIN-RETORTILLO, L. "Los conceptos de consolidación, rehabilitación y restauración en la Ley de Patrimonio Histórico español" en El Derecho Administrativo en el umbral del siglo XXI. Libro Homenaje al Profesor Dr. D. Ramón Martín Mateo, Tirant lo Blanch, Valencia, 2000, p. 3193.

MARTÍNEZ ALOY, J. "Provincia de Valencia" en Geografía General del Reino de Valencia, ed. Alberto Martín, Barcelona, 1988, pp.150–250.

MARTINEZ GARCIA "Inventario y catalogación del Patrimonio Histórico. Andalucía como marco" in Patrimonio Cultural y Derecho **7**, 2003.

MAYES, T. "Impact Assessments as a Tool for Protection against the State". In Legal and Financial Aspects of Architectural Conservation: The Smolenice Castle Conference. Dundurn Press, 1997.

MCLAMB, S. Preservation law survey 2001: State preservation law in 8 Wid. L. Symp. J. 463.

MILLER, J. "A Layperson's Guide to Historic Preservation Law: A Survey of Federal, State, and Local Laws Governing Historic Resource Protection", in 1 Historic Preservation Law 3 (ALI-ABA Committee on Continuing Professional Education, 1999).

MORALES MIRANDA, J. "La interpretación del patrimonio natural y cultural: todo un camino por recorrer", Boletín del Instituto Andaluz del Patrimonio Histórico **25**, 1998, pp. 150–157.

MORALES MIRANDA, J. Guía práctica para la interpretación del patrimonio: el arte de acercar el legado natural y cultural al público visitante, Junta de Andalucía, Sevilla, 1998.

MORENO NAVARRO, A. "El antropólogo y la defensa del patrimonio cultural", Actas del 2º Congreso de Antropología, Madrid, pp. 53–55.

MUÑOZ MACHADO, S. La resurrección de las ruinas. El caso del Teatro Romano de Sagunto, Civitas, Madrid, 2002.

MUÑOZ MACHADO, S. Derecho Público de las Comunidades Autónomas, vol. I, Madrid, Civitas, 1982.

MURTAGH, W. Keeping Time: The History and Theory of Preservation in America. Pittstown, NJ: Main Street Press, 1988.

MYNORS, C. Listed Buildings, Conservation Areas and Monuments, Sweet and Maxwell, 2006.

MYNORS, C. The Law of Trees, Forests and Hedgerows, Sweet & Maxwell, London, 2002.

NIVALA, J. "Saving the Spirit of Our Places: A View on Our Built Environment" 15 UCLA J. Environmental. Law & Policy **1**.

OFFICE OF THE DEPUTY PRIME MINISTER Our Towns and Cities: The Future – Delivering an Urban Renaissance, November 2000.

OFFICE OF THE DEPUTY PRIME MINISTER Planning Policy Guidance 15: Planning and the historic environment, September 1994.

OFFICE OF THE DEPUTY PRIME MINISTER Review of Permitted Development Rights: 2003.

OROZCO PARDO, G. & PEREZ ALONSO, E. La tutela civil y penal del Patrimonio Histórico, cultural y artístico, ed. McGraw Hill, Madrid, 1996.

PALMER, M. & NEAVERSON, P. Industrial Archaeology: Principles and Practice, 1998.

PALMER, M. & NEAVERSON, P. Industry in the Landscape, 1700–1900, 1994

PAPADEMETRIOU KOURTIS, J. "The Constructive Trust: Equity's Answer to the Need for a Strong Deterrent to the Destruction of Historic Landmarks", 16 B.C. Envtl. Aff. L. Rev., pp. 793, 804, 1989.

PAREJO ALFONSO, L. "Urbanismo y patrimonio histórico", Revista de Patrimonio Cultural y Derecho **2**, 1998, pp. 55–79.

PÉREZ ANDRÉS, A. La ordenación del territorio en el Estado de las Autonomías, Marcial Pons, 1998.

PÉREZ DE ARMIÑAN YDE LA SERNA, A. "Una década de aplicación de la ley del patrimonio histórico español", Patrimonio Cultural y Derecho **1**, 1997, pp. 33–51.

PEREZ LUÑO, A. Los Derechos fundamentales, Tecnos, 7ª edición, Madrid, 1998.

PEREZ LUÑO, A. (Cord.), Derechos humanos y constitucionalismo ante el tercer milenio, Marcial Pons, Madrid, 1996.

PÉREZ QUESADA, P. "Análisis del valor y evaluación multicriterio en la gestión del patrimonio histórico", Boletín del Instituto Andaluz del Patrimonio Histórico **28**, 1999, pp. 56–71.

PHELAN, M. "A Synopsis of the Laws Protecting Our Cultural Heritage" in 28 New Eng.L. Rev. **63** Fall 1993, p. 63 y ss.

PORRO GUTIÉRREZ, J.M. "Patrimonio y cultura: dos términos en interacción", Boletín del Instituto Andaluz del Patrimonio Histórico **12**, 1995, pp. 27–37.

PRADA BENGOA, J.I. "La protección del patrimonio cultural de la humanidad", Boletín del Instituto Andaluz del Patrimonio Histórico **17**, 1996, pp. 62–71.

PRENTICE, R. Tourism and Heritage places, Routledge, 1993. See also BONIFACE P. & FOWLER P.J. Heritage and Tourism in the global village', London: Routledge, 1993.

PROTT, LYNDEL V. & O'KEEFE, P.J. Law and the cultural heritage Professional Books, 1984.

RAND, K. "Nothing lasts forever: toward a coherent theory in American preservation law" in 27 University of Michigan Journal of Law Reform 277 1993.

RAUSELL KÖSTER, P. & CARRASCO ARROYO "El Patrimoni Industrial de Sagunt: un factor posible de desenvolupament territorial", Universidad de Valencia, Valencia, 1999.

REVERT ROLDAN, X. & MONTIEL ROIG, G. Presentación al libro Reconversión y Revolución, Industrialización y Patrimonio, edit. Universidad de Valencia, 1999, pp. 10–50.

REVERT ROLDAN, X. "Arte industrial, patrimonio y acción social: los Altos Hornos del Mediterráneo en Puerto de Sagunto", Boletín del Instituto Andaluz del Patrimonio Histórico, number 21, 1997, pp. 112–117.

RIOJA LÓPEZ, C. "Reflexiones en torno a la cultura inmaterial y su gestión patrimonial en la Comunidad Autónoma andaluza". En: Boletín del Instituto Andaluz del Patrimonio Histórico **16**, 1996, pp. 79–84.

RODRÍGUEZ BECERRA, S. "Patrimonio cultural, patrimonio antropologico y museos de antropología", Boletín del Instituto Andaluz del Patrimonio Histórico **21**, 1997, pp. 42–52.

RODRÍGUEZ CUADROS, E. "Los hijos de Prometeo: historia de un negro suelo amado" en Reconversión y Revolución, Industrialización y Patrimonio, edit. Universidad de Valencia, 1999.

RODRÍGUEZ NUÑEZ, A. "La protección del patrimonio cultural en el Código Penal de 1995", Patrimonio Cultural y Derecho **2**, 1998, pp. 133–161.

RODRÍGUEZ OLIVA, M.C. "Información de bienes culturales. Sistemas de información de actividades de gestión patrimonial (S.I.A.G.P.) Plan general de bienes culturales 1989–1995", Boletín del Instituto Andaluz del Patrimonio Histórico **19**, 1997, pp. 118–127.

RODWELL, D. Conservation and Sustainability in Historic Cities, Blackwell Publishing, 2007, p. 12

ROSE, R. Preservation and Community: New Directions in the Law of Historic Preservation, Stan. L. Rev. **33**, 1981, pp. 473, 497–502.

ROSE, R. Artistic Heritage in a changing Pacific, University of Hawaii, 1993.

RÖSSLER, M. World Heritage cultural landscapes, The George Wright Forum, The Journal of the George Wright Society **17(**1), pp. 27–34. (Special issue: Landscape Stewardship. New Directions in Conservation of Nature and Culture) 2000.

ROTHSTEIN, S. "Takings Jurisprudence Comes In From the Cold: Preserving Interiors through Landmark Designation" in Conn. L. Rev. **26**, p. 1105.

RUÍZ GIL, J.A. "Patrimonio cultural y desarrollo local: estudio de casos de patrimonio arqueológico en ámbitos rurales", Boletín del Instituto Andaluz del Patrimonio Histórico **18**, 1997, pp. 38–44.

RUIZ RICO, G. "La disciplina constitucional del Patrimonio Histórico en España", Patrimonio cultural y Derecho **4**, 2000, pp. 29–65.

SALAS ÁLVAREZ, J.A. "Directrices comunitarias para la restitución de Bienes Culturales ilícitamente exportados", Boletín del Instituto Andaluz del Patrimonio Histórico **8**, 1994, pp. 32–33.

SÁNCHEZ–PALENCIA RAMOS, F.J. "Reflexiones sobre la nueva Ley del Patrimonio Histórico Español y la Arqueología", AIC: Análisis e Investigaciones Culturales **25**, 1985, pp. 13–19.

SANZ NAJERA, M.S. "La conservación del Patrimonio Mueble según la Ley 13/1985 de 25 de junio del Patrimonio Histórico Español", AIC: Análisis e Investigaciones Culturales **25**, 1985, pp. 23–27.

SCHOEN, R. "Confronting the Appalachian breakdown: historic preservation law in Appalachia and the potential benefits of historic preservation for rural communities" in W. Va. L. Rev. **110** (Spring), 2008, p. 1303.

SCHOUTEN, F. "Heritage as a Historical Reality" in Heritage Tourism and Society, ed. HERBERT, T., Pinter Press, 1995.

SIEGEL, S. A Student's Guide to easements, Real Covenants, and Equitable Servitudes, 1988.

SOLIAS I ARIS, J.M. "Reflexions sobre arqueologia industrial",Afers: fulls de recerca i pensament **8**(1988–1989), pp. 527–535.

SQUILLACE, M. "The Monumental legacy of the antiquities act of 1906" in Ga. L. Rev. **37** (Winter), p. 473.

STRATTON, M. & TRINDER, B. Twentieth Century Industrial Archaeology, 2000.

SWENSON, R. "History as a Catalyst for University Involvement in Regional Historical Research, Regional Planning, and Community/Economic Development." P.A.S.T. Pioneer Society Transactions **26**, 2003, pp. 46–55.

THERIVEL, R. Strategic Environmental Assessment in Action, Earthscan 2004.

THOMAS, K. Development Control: Principles and Practice Routledge Press, 1997.

TILEY, J. Revenue Law, Hart Publishing, 2005.

TRINDER, B. (ed.) The Blackwell Encyclopaedia of Industrial Archaeology, 1992.

VALLE PÉREZ, M., PÉREZ HUMANES, M.A., TAPIA MARTÍN, C. & VARONA, M. "Desarrollo de la base de datos inmuebles. Algunas reflexiones sobre el destino de la documentación patrimonial", Boletín del Instituto Andaluz del Patrimonio Histórico **19**, 1997, pp. 131–137.

VALVERDE MERINO, J.L. "Legislación autonómica sobre el patrimonio histórico y museos", Revista de Museología **14**, 1998, pp. 117–123.

VERA BOTÍ, A. "El patrimonio urbanístico y arquitectónico: de la ambivalencia a la crisis", Arquimur **5**, 1988, pp. 74–79.

VERDUGO SANTOS, J. "El uso de los fondos estructurales europeos: la recuperación del patrimonio histórico en el marco de las industrias alternativas", Boletín del Instituto Andaluz del Patrimonio Histórico **7**, 1994, pp. 32–35.

VIÑAS TORNER, V. "La conservación del Patrimonio Bibliográfico y Documental según la Ley 13/1985, de 25 de junio, del Patrimonio Histórico Español", AIC: Análisis e Investigaciones Culturales **25**, 1985, pp. 23–27.

ZAMORA BAÑO, F. "Ética y patrimonio cultural" Boletín del Instituto Andaluz del Patrimonio Histórico **21**, 1997, pp. 37–41.

ZAX, L. Protection of the Built Environment: A Washington, D.C. Case Study in Historic Preservation, B.C. Envtl. Aff. L. Rev. **19**, 1992, p. 651.